SpringerBriefs in Energy

For further volumes:
http://www.springer.com/series/8903

Rita Pandey · Sanjay Bali · Nandita Mongia

The National Clean Energy Fund of India

A Framework for Promoting Effective Utilization

 Springer

Rita Pandey
National Institute of Public Finance
   and Policy (NIPFP)
New Delhi
India

Sanjay Bali
New Delhi
India

Nandita Mongia
Project Design and Monitoring, Energy
   and Climate Change
Dwarka
Delhi
India

ISSN 2191-5520   ISSN 2191-5539   (electronic)
ISBN 978-81-322-1963-7   ISBN 978-81-322-1964-4   (eBook)
DOI 10.1007/978-81-322-1964-4
Springer New Delhi Heidelberg New York Dordrecht London

Library of Congress Control Number: 2014942068

Printed on acid-free paper

Springer is part of Springer Science+Business Media (www.springer.com)

Foreword

The National Clean Energy Fund (NCEF), announced in Budget 2010–2011, is seen as a major step in India's quest for energy security and reducing carbon intensity of energy. Funding research and innovative projects in clean energy technologies, and harnessing renewable energy sources to reduce dependence on fossil fuels constitute the objectives of the NCEF. It is observed that utilization of funds from NCEF has been rather low and disbursements, so far, are aligned more with ongoing programs/missions of various ministries/departments than with the stated objectives of the fund. This poses potential risk of diluting the focus of NCEF with adverse implications for research and innovation in clean energy sector in India. Especially, in the absence of any identified targets and prioritization.

This study aims to provide a detailed framework for promoting effective utilization and administration of NCEF. It is hoped that the recommendations of the study will inform the government so that appropriate corrections may be made timely. The outputs of the study will also be useful to hone the strategic thinking on a suitable energy technology policy and an assessment of technology needs besides other barriers in clean energy sector in India.

Rathin Roy

Acknowledgments

This study has benefitted from the valuable suggestions received from experts in government, industry, and academia. In particular, we are grateful to Meena Agarwal (Joint Secretary, Ministry of Finance, Government of India), Saurabh Garg (Joint Secretary, Ministry of Finance, Government of India), P. R. Shukla (Professor, IIM Ahmedabad), Anil Gupta (Professor, IIM Ahmedabad), U. Sankar (Professor Emeritus, Madras School of Economics, Chennai), Shailly Kedia (Associate Fellow, TERI), and Bipin Shah (Professor, Entrepreneurship Development Institute of India, Ahmedabad) for insights shared on this topic.

We acknowledge the feedback and constructive ideas received from a number of stakeholders including the participants of the two workshops held in November, 2012, and April, 2013 in Delhi. We are extremely grateful to panellists—Kirit Parikh (Chairman, IRADe), Anil K. Jain (Advisor, Planning Commission), Meena Agarwal (Joint Secretary, MoF), D. N. Prasad (Advisor, MoC), Rakesh Bhalla (Advisor, IREDA), K. Usha Rao (Senior Project Manager, KfW), and Pradeep Dadhich (Director, Deloitte India) of the workshops for their valuable inputs. A list of experts and stakeholders who were consulted during the course of the study is included in Annexure 1.

Research assistance by Sargam Gupta, Lipi Budhraja, and Nitesh Khandelwal is gratefully acknowledged. Wasim Ahmad provided secretarial assistance.

We would also like to thank Shakti Foundation for the opportunity to undertake this study and extending financial support which made this study possible.

Rita Pandey

Contents

Figures

Tables

Abbreviations

ACF	Australian Conservation Foundation
BEE	Bureau of Energy Efficiency
CalCEF	California Clean Energy Fund
CBGA	Centre for Budget and Governance Accountability
CBEC	Central Board of Excise and Customs
CCEA	Cabinet Committee on Economic Affairs
CCEF	Connecticut Clean Energy Fund
CCS	Carbon Capture and Storage
CDM	Clean Development Mechanism
CEA	Central Electricity Authority
CEFC	Clean Energy Finance Corporation
CERC	Central Electricity Regulatory Commission
CII	Connecticut Innovations Incorporated
CMPDIL	Central Mine Planning and Design Institute Limited
COMAPS	Coastal Ocean Monitoring and Prediction Systems
CSLF	Carbon Sequestration Leadership Forum
CSR	Corporate Social Responsibility
DEDE	Department of Alternative Energy Development and Efficiency
DSM	Demand Side Management
DST	Department of Science and Technology
ECA	Energy Conservation Act 2001
ENCON	Energy Conservation Promotion Fund
EPPO	Energy Policy and Planning Office
FBC	Fluidized Bed Technology
FCM	Federation of Canadian Municipalities
GDP	Gross Domestic Production
GEF	Global Environment Facility
GMF	Green Municipal Fund
ICMAM	Integrated Coastal and Marine Area Management
IEP	Integrated Energy Policy
IMG	Inter-Ministerial Group
INCCA	Indian Network for Climate Change Assessment
IPPs	Independent Power Producers

IREDA	Indian Renewable Energy Development Agency
IREP	Integrated Rural Energy Program
LFA	Logical Framework Approach
LOICZ	Land Ocean Interactions in the Coastal Zone
M&E	Monitoring and Evaluation
MESITA	Malaysian Electricity Supply Industries Trust Account
MoF	Ministry of Finance
MNRE	Ministry of New and Renewable Energy
MTC	Massachusetts Technology Collaborative
NABARD	National Bank for Agriculture and Rural Development
NAPCC	National Action Plan on Climate Change
NCEF	National Clean Energy Fund
NEDO	New Energy and Industrial Technology Organization
NEP	National Electricity Policy
NEPC	National Energy Policy Council
NGO	Non-Government Organization
NIF	National Innovation Foundation
NJCEP	New Jersey's Clean Energy Program
NMEEE	National Mission for Enhanced Energy Efficiency
NMSA	National Mission for Sustainable Agriculture
NMSKCC	National Mission on Strategic Knowledge for Climate Change
NSDC	National Skill Development Corporation
PAT	Perform, Achieve and Trade
PFC	Power Finance Corporation
PG & E	Pacific Gas and Electric Company
PGCIL	Power Grid Corporation India Limited
REC	Renewable Energy Certificate
ROtI	Review of Outcomes to Impacts
RPO	Renewable Purchase Obligation
SHP	Small Hydro Power
SICOM	Society of Integrated Coastal Management
SMART	Specific, Measurable, Attainable, Relevant and Tractable
SMEs	Small and Medium Enterprises
TDDP	Technology Development and Demonstration Program
TNB	Tenaga Nasional Berhad
TP	Tariff Policy
UNEP	United Nations Environment Programme
UNFCCC	United Nations Framework Convention on Climate Change

About the Authors

Rita Pandey is a Professor at the National Institute of Public Finance and Policy, New Delhi. She obtained her Ph.D. in Economics from the Indian Institute of Technology Kanpur and was a visiting fellow at the School of Forestry and Environmental Studies, University of Yale, USA, in 2001. She has been a member of the Steering Committee to implement the Montreal Protocol, appointed by the Ministry of Environment and Forests, Government of India; and a member of the Indo-German Expert Group on Green and Inclusive Economy. Dr. Pandey's research interests span integrating environmental, and natural resource concerns into policy analysis focusing on energy, industry, climate change, and urban transport through the use of environmental economics, and tools and measures of public finance. She has undertaken a broad range of studies examining the different links between environment and resources and the economy and has worked extensively on economic instruments and environmental fiscal reforms for environmental management and resource conservation. She has published widely in national and international journals and has four books to her credit.

Sanjay Bali is currently an independent consultant, with interests in many areas of research, including innovation, deployment, and dissemination of new energy and clean technologies. He has been involved in projects covering technological and policy aspects of new energy and clean technologies. In the past 18 years. Mr. Bali has been associated, in various capacities, with organizations such as The Energy and Resources Institute, New Delhi; InsPIRE Network for Environment, New Delhi; Deutsche Gesellschaft für Internationale Zusammenarbeit (GIZ), New Delhi; The Department of International Development, UK (DfiD), New Delhi; Asian and Pacific Centre for Transfer of Technology, New Delhi and National Institute of Public Finance and Policy, New Delhi. He has a Masters in Chemistry from the Indian Institute of Technology, Delhi.

Nandita Mongia is a freelance consultant in the field of energy and climate change mitigation. With a background in academia and extensive international development experiences, she holds a Ph.D. in Optimal Technology Choice and Production Planning in coal Industry. She is a Fulbright Postdoctoral Fellow and been a visiting staff scientist at the Lawrence Berkeley National Laboratory, California, USA. She has been a resource person for the US Country Studies Program on Energy—Technology

Choice Options model for CO_2 emission reduction, to the Science Technology and Environment Program in Princeton, East-West Centre, Hawaii, United Nations University-Tokyo-Japan, Energy Research Institute, Beijing, China. Dr. Mongia has been interviewed by international print and visual media on issues of clean energy, energy access, climate change, constraints in sustainable development, etc. She has several publications to her credit.

Abstract

The need for energy security in India stems mainly from the need to maintain the growth trajectory necessary to achieve its objectives of poverty eradication and inclusive development. In this context, it is being increasingly recognized that going forward, the country needs to diversify its primary energy sources and make attempts to explore cleaner and renewable sources of, and solutions for energy. The arguments are: the need to ensure energy security by reducing dependence on fuel imports, securing development dividends through poverty linkages, GHG emissions and the risk of climate change, and health.

A number of financial, market-based, and regulatory measures have been put in place toward this end. However, sustainable and successful deployment and adoption of clean energy solutions are not a financial or a regulatory issue alone, it has a direct bearing on the need and capacity for research and innovation, skill development, besides addressing other market barriers including many social and cultural issues.

In this context, setting up of the National Clean Energy Fund (NCEF) for funding research and innovative projects in clean energy technologies is a welcome step. The present design of NCEF lacks robust strategy, targets, and innovative ideas on prioritization. In this context, this study, among others, identifies the most promising avenues for utilization of the Fund's resources and their distribution across the energy subsectors; examines catalytic opportunities for the Fund, develops a framework for evaluation of the Fund's performance as well as a framework for assessment of proposals and monitoring and evaluation of projects funded by the NCEF; and makes specific recommendations on addressing other crucial market barriers.

This book will be of use to policy makers, government agencies, academics, industries, and other stakeholders working in the field of clean energy and energy technology policy.

Keywords Designing clean energy fund · Innovation · Energy policy · Clean energy technology policy · Project evaluation protocol · Clean energy fund evaluation protocol · Barriers in clean energy industry · Clean energy financing mechanisms

Chapter 1
Context and Objectives of the Study

1.1 Background

India is one among the fastest growing economies in the world. This growth is dependent on energy, so maintaining this growth trajectory would require the country to ensure its energy security. This is important for India not only to support its economic growth but also to achieve its objectives of poverty eradication and inclusive development.

The Indian government is aware of the magnitude and importance of the challenges involved, at the same time realizes its responsibility and voluntary commitment to combating climate change. It is being increasingly recognized that going forward the country needs to diversify its primary energy sources and attempt to explore cleaner and renewable sources of and solutions for energy.[1] The arguments are the need to ensure energy security by reducing dependence on fuel imports, securing development dividends through poverty linkages, GHG emissions

---

[1] The term *clean energy* typically refers to renewable and non-polluting energy sources. Renewable energy is derived from natural resources that can be replenished constantly. Renewable energy takes various forms and includes electricity and heat generated from solar, wind, ocean, hydropower, biomass, geothermal resources, and biofuels and hydrogen-derived from renewable resources. In addition, certain clean coal technologies and energy efficiency measures also fall under the broad definition of clean energy initiatives. The term *clean energy solutions* broadly refers to systems which promote, enhance or advance the energy generation, transport, storage and use so as to reduce the environmental footprint and decrease energy intensity. In the context of the debate on global warming, environmental footprint is typically measured as carbon footprint. Such systems include products, services, technologies, and regulatory and market-based incentives. These have typically focused on the six key sectors: power, transport, industry, buildings, carbon sequestration, and carbon capture and storage.

R. Pandey et al., *The National Clean Energy Fund of India*, SpringerBriefs in Energy, DOI: 10.1007/978-81-322-1964-4_1, © The Author(s) 2014

and the risk of climate change, and health benefits of cleaner and renewable energy and clean energy solutions.

For a developing country like India it is a daunting task. What is encouraging, however, is that it brings potentially huge opportunities for economic growth and employment generation and gains from trade in ever-growing international market for energy. Gainful exploitation of these, however, would require a clear vision and multipronged approach.

India has taken several important measures and has made a steady progress by putting in place a number of institutions, mechanisms and policies, although a lot remains. In this context, a recent report of a high-level Expert Committee on Integrated Energy Policy (IEP 2006) is an important step. Further, to address climate change issues in energy, India has announced a domestic goal of reducing the emission intensity of its GDP by 20–25 % of the 2005 level by 2020 which would require sector-specific actions involving substantial financial outlay, technology choices and research and innovation.

The eight National Missions which form the core of the National Action Plan on Climate Change (NAPCC) adopted in 2008 have both mitigation and adaptation measures. While adaptation is the focus of the NAPCC, missions on solar energy and energy efficiency are geared to mitigation. Apart from the NAPCC, all the states are in different stages of preparing state-level action plans. These plans are envisioned as extensions of the NAPCC at various levels of governance, aligned with the eight National Missions.

The major policies and actions in addressing energy security, climate change mitigation and adaptation cut across different sectors and areas of the economy. The initiatives in some of the major areas are as follows.

1.1.1 National Clean Energy Fund

The National Clean Energy Fund (NCEF), announced in Budget 2010–2011 (MOF 2010), is seen as a major step in India's quest for energy security and reducing carbon intensity of energy. The Union Finance Minister in his budget speech in Parliament said that "There are many areas of the country where pollution levels have reached alarming proportions. While we must ensure that the principle of 'polluter pays' remains the basic guiding criteria for pollution management, we must also give a positive thrust to development of clean energy. I propose to establish a National Clean Energy Fund for funding research and innovative projects in clean energy technologies (Paragraph 66)... Harnessing renewable energy sources to reduce dependence on fossil fuels is now recognized as a credible strategy for combating global warming and climate change. To build the corpus of the National Clean Energy Fund, I propose to levy a clean energy cess on coal produced in India at a nominal rate of Rs. 50 per tonne. This cess will also apply to imported coal (Paragraph 154)".

1.1.2 Energy Efficiency

The Energy Conservation Act 2001 (ECA 2001) empowers the government to prescribe and ensure compliance with standards and norms for energy consumers, and prescribe energy conservation building codes and energy audits. Apart from these, there are a range of programmes being implemented by the Bureau of Energy Efficiency (BEE) in key sectors of energy demand.

Among the recent initiatives, the National Mission for Enhanced Energy Efficiency (NMEEE) is the key focus for government action for energy efficiency. The NMEEE is divided into four components: (a) perform, achieve and trade (PAT), a scheme for trading in energy efficiency certificates that will cover about 700 industrial units and achieve a saving of almost 17,000 MW of energy by 2017. This scheme is mandatory for all large industrial units and facilities in thermal power, aluminium, cement, fertilizers, chlor-alkali, steel, paper and pulp and textiles; (b) energy efficiency financing platform; (c) market transformation for energy efficiency; and (d) framework for energy efficient economic development.

1.1.3 Power Plants

For reducing emission intensity, 60 % of coal-based capacity addition in the Twelfth Plan and 100 % in the Thirteenth Plan shall be done by deploying supercritical technology. Ultra-supercritical power plants operate at higher efficiency. The first ultra-supercritical power plant is expected in 2017. Large-scale adoption of this technology after a few years would further reduce the emission intensity of the Indian power sector. Also, there are plans to retire old and inefficient coal-based power generating units.

1.1.4 Renewable Energy

The Electricity Act 2003 together with the National Electricity Policy 2005 (NEP 2005) and the Tariff Policy (TP) envisage regulatory interventions for promotion of renewable energy sources. In this context, the initiatives of the Central Electricity Regulatory Commission (CERC) include determination of preferential tariff for renewable energy, creating a facilitative framework of grid connectivity through the Indian Electricity Grid Code and developing market-based instruments such as Renewable Energy Certificate (REC). The REC mechanism is seen as a major market-based initiative towards promoting renewable energy and encouraging competition in this segment. It addresses the twin objectives of harnessing renewable energy sources in areas with high potential and compliance with Renewable Purchase Obligation (RPO) by resource-deficit states.

1.1.5 Nuclear Energy

India recognizes the importance of nuclear energy as a sustainable energy source. Present nuclear-installed capacity is 4,780 MW, and there are plans to increase the generation capacity to 20,000 MW by 2020.

1.1.6 Transport

India has taken substantial initiatives to make the transport sector less emission intensive. One of the major initiatives has been up-gradation of vehicular emission norms. The commercial manufacture of battery-operated vehicles has begun in India. In addition to this, Integrated Transport Policy 2001 (ITP 2001) promotes the use of ethanol-blended petrol and biodiesel. The National Urban Transport Policy emphasizes the development and usage of extensive public transport facilities (including non-motorized modes) over personal vehicles. Besides, there has been a large-scale switchover from petrol and diesel to CNG.

1.1.7 Agriculture and Forestry

One of the recent and key policy initiatives is National Mission for Sustainable Agriculture (NMSA). In addition, there are programmes for crop improvement and drought proofing. India has launched an ambitious Green India Mission to increase the quality and quantity of forest cover in 10 million hectare. of land. Also, an incentive-based additional special grant of about Rs. 6,500 crores had been announced by the central government to all states for sustainable forestry management. Other policies and programmes in the forestry sector include the National Forest Policy (1988), Participatory Forest Management/Joint Forest Management Program, National Afforestation Program, National Forestry Action Program and National Watershed Development Project for rain-fed areas.

1.1.8 Marine and Coastal Environment

Ensuring stability in the coastal environment in India becomes imperative considering its densely inhabited, long coastline of more than 7,500 km. Some of the major initiatives taken in this area are Coastal Ocean Monitoring and Prediction Systems (COMAPS), Land Ocean Interactions in the Coastal Zone (LOICZ), Integrated Coastal and Marine Area Management (ICMAM) and Society of Integrated Coastal Management (SICOM).

1.1.9 Initiatives for Enhancing Knowledge and Scientific Findings

Besides the National Mission on Strategic Knowledge for Climate Change, a network, Indian Network for Climate Change Assessment (INCCA), has been set up to carry out scientific studies of various aspects of climate change. The INCCA has recently carried out a 4 × 4 assessment of climate change in India covering four major sectors in four ecological regions of the country and an updated inventory of the GHG emissions for the year 2007.

1.1.10 Enhancing Adaptive Capacity

India's strategy for enhancing its adaptive capacity to climate variability is reflected in many of its social and economic development programmes. Several of India's social-sector schemes, with their emphasis on livelihood security and welfare of the weaker sections, aim to empower them to cope with uncertainties in the long run. India implements a series of central sector and centrally sponsored schemes under different ministries/departments aimed at achieving social and economic development. Many of these schemes have substantial climate change adaptation orientation. An exercise has been carried out to measure the expenditure on adaptation-related programmes with critical adaptation components: (a) crop improvement and research, (b) poverty alleviation and livelihood preservation, (c) drought proofing and flood control, (d) risk financing, (e) forest conservation, (f) health and (g) rural education and infrastructure. Estimates show that India's expenditure on these adaptation-oriented schemes has increased from 1.45 % of GDP in 2000–2001 to 2.82 % during 2009–2010. This is a reflection of the multiplicity of economic and social welfare programmes under implementation in India.

1.2 Objectives of the Study

India's efforts towards achieving energy security, social inclusion and environmental targets—reduction in emission intensity of energy—can be categorized into (1) energy efficiency and conservation (both demand and supply side) to cut carbon and other emissions besides resource conservation; (2) switch to renewable energy to reduce the share of fossil fuel-based energy; and (3) land-use changes and forestry as a net sink of carbon. A number of financial, market-based and regulatory measures have been put in place towards this end.

However, sustainable and successful deployment and adoption of clean energy solutions are not a financial or a regulatory issue alone, but also have a direct bearing on the need and capacity for research and innovation, and skill development, besides addressing other market barriers.

Although India has built up significant technological and innovation capacity since independence in many areas including chemicals, pharmaceuticals, information technology, atomic energy and space technology, there is little focus on strategic planning for and promotion of research and innovation in the energy sector both in terms of expenditure and institutional support. *Not only are these relatively small but also fragmented to make the desired impact.* There also exists, in general, a lack of technical expertise in installation, operations, maintenance, troubleshooting and other aspects of implementation of clean energy.

India needs a dedicated energy technology policy and planning body and a roadmap to finance identified options. Once this is in place, other policies and regulations can pitch in to facilitate favourable market conditions necessary for success and sustainability of the chosen path.

This will help not only to develop products, devices and processes appropriate for India in terms of price, performance requirement, raw material suitability and other needs and constraints, but will also open up opportunities to gain from being a participant in the global market.

The setting up of the NCEF for funding research and innovative projects in clean energy technologies is a welcome step. Although the objectives of NCEF are broadly in line with the critical needs of the clean energy sector in India, it lacks guidance on the overall vision and strategy that will be necessary to realize its objectives thus emphasizing the need for a critical review of the present framework and operation of NCEF with a view to identifying appropriate measures that would help promote its effective utilization. In this context, specific objectives of this study are as follows:

1. A review of existing framework of the NCEF including its structure and proposed mode of administration, to identify the key gaps in the existing framework in light of its stated objectives.
2. An assessment of the structure and workings of similar funds to identify the principles and best practices that are applicable to, and where relevant, their adoption can improve the functioning of the NCEF.
3. Identify the most promising avenues for utilization of the Fund's resources, given its stated objectives. The key questions asked and investigated in this domain include the following:
   (a) Given the nature and size of the fund, should resources be directed towards specific energy sub-sectors? If so, how should these sub-sectors be selected?
   (b) What type of projects should be supported by the fund? Should the emphasis be on using NCEF resources for catalytic opportunities, such as for establishing institutions and leveraging private capital, or should the focus be on promoting deployment of new technologies by financing projects that result in on-the-ground creation of new generation facilities based on clean energy sources such as solar and wind?

4. Develop a multi-criteria-based framework that can be used for evaluating the NCEF's performance over time.
5. Develop a project level evaluation protocol/multi-criteria framework for the following:
   (a) Assessment of proposals submitted for funding through the NCEF.
   (b) Monitoring and evaluation of individual projects funded by the NCEF.
6. Based on the above analyses develop a set of recommendations that will promote effective utilisation of the NCEF.

References

ECA (2001) Energy conservation act, Ministry of Power, Government of India
IEP (2006) Integrated energy policy: report of Expert Committee, Planning Commission, Government of India
MOF (2010) Union budget 2010–2011, Ministry of Finance, Government of India
NEP (2005) National electricity policy, Ministry of Power, Government of India
ITP (2001) Report of the task force on integrated transport policy, Planning Commission, Government of India

Chapter 2
Existing Framework and Operation of NCEF: A Review

Objective of this chapter was to critically review the existing framework of NCEF, including its structure and proposed mode of administration, and identify the key gaps. The scope of this exercise has been limited by the identified objectives of the NCEF.

2.1 Existing Framework of NCEF

Subsequent to the announcement of setting up of NCEF, the Central Board of Excise and Customs (CBEC) issued a notification dated 22 June 2010 (CBEC 2010) to notify the Clean Energy Cess Rules, 2010.[1] In 2011, the Cabinet Committee on Economic Affairs (CCEA) approved the constitution of the NCEF under the Public Accounts of India along with the guidelines and modalities for approval of projects to be funded from the Fund.

The Fund has been set up to serve as a separate non-lapsable corpus. Plan Finance II Division of the Department of Expenditure, Ministry of Finance (MoF), Government of India, is the nodal agency for administering the Fund and has drafted the Cabinet note outlining the framework of the NCEF. In April 2011, the MoF issued the approved guidelines for appraisal and approval of the project/schemes eligible for funding under the NCEF along with an indicative list of such projects (MoF 2011a). Subsequently, in June 2011, the MoF issued a format to invite proposals under the NCEF for consideration (MoF 2011b). A summary of the main points in these documents outlining the objectives of the NCEF and salient features of how the Fund will be operationalized is as follows.

---

[1] The cess is levied as a duty of excise on coal, lignite and peat. It applies to the gross quantity of these raw materials raised and dispatched from a coal mine except on coal produced in Meghalaya. No deduction from this quantity is allowed for loss on account of washing of coal or its conversion into any other product or form prior to its dispatch from the mine. To avoid double levy, the cess is not chargeable on washed coal or any other form. Imported coal, including washed coal, also attracts cess in the form of additional duty of Customs.

R. Pandey et al., *The National Clean Energy Fund of India*, SpringerBriefs in Energy, DOI: 10.1007/978-81-322-1964-4_2, © The Author(s) 2014

2.2 Objectives of NCEF

As per NCEF guidelines, "The NCEF is created for funding research and innovative projects in clean energy technologies. Any project/scheme relating to innovative methods to adopt to clean energy technology and research and development shall be eligible for funding under the NCEF".

> While the objectives of NCEF seem to be in line with the critical needs of the clean energy sector in India, there is no guidance on the overall vision and the strategy that will be employed to realize these.

2.3 Projects Eligible for Funding Under NCEF

Para 2.1 of the NCEF guidelines provides an indicative list of projects eligible for funding.

These can be grouped broadly into the following categories:

- Advanced technologies in clean fossil energy.
- Advanced technologies in renewable energy including critical energy evacuation infrastructure, and integrated community energy solutions.
- Basic energy sciences.
- Projects related to environment management particularly in geographical areas surrounding the energy sector projects.
- Pilot and demonstration projects for commercialization.
- Projects identified in NAPCC and those relating to R&D to replace existing technologies under national mission on Strategic Knowledge for Climate Change (NMSKCC).

> This list however, is too broad based and appears to encompass every possible action required to cope with climate change. This poses potential risk of diluting the focus of NCEF with adverse implications for research and innovation in clean energy sector in India. Especially so, in the absence of any identified targets and prioritisation.

2.4 Mode of Appraisal and Approval of Project Proposals

NCEF guidelines outline the process of appraisal of project proposals received for consideration for funding from the NCEF as follows:

- The project proposals can be submitted for seeking NCEF support only through a relevant central ministry/department. The first examination of the proposal is done at this level, and if deemed fit, the proposal is forwarded for comments to the Planning Commission, MoF, and any other relevant ministry for review and comments. The third and final review is done by an Inter-Ministerial Group (IMG) which has been constituted to appraise the projects/schemes and make recommendations for approval.

- The IMG comprises:
    - Finance Secretary, MoF-Chairperson
    - Secretary (Expenditure), MoF
    - Secretary (Revenue), MoF
    - Principal Scientific Advisor to the Government of India
    - Representative of Planning Commission
    - Representatives of Ministry sponsoring the proposal and other Ministries concerned with that specific proposal.

- While projects under Rs. 150 crore can be approved by the Minister incharge of the project sponsoring/line ministry/department, projects of Rs. 150 crore and under 300 crore will need approval from both the Minister incharge of the project sponsoring/line ministry/department as well as the Finance Minister. Projects of Rs. 300 crore and above will require approval from the CCEA.
- The IMG may seek the assistance and views of technical experts from related organisations and individuals of repute in the area of clean energy to review, evaluate and recommend projects.
- To monitor the progress of the NCEF funded projects, the IMG will identify/ appoint appropriate professional agencies.
- There will be a time frame specified under the scheme for processing of applications at each stage.

The requirement to apply through a Central government Ministry/department is faulty. Window for direct application should be there. Given the objectives of the Fund a dedicated team/mission will be required to administer it. The present structure does not seem adequate and the most appropriate.

2.5 Funding Limit, Eligibility and Funding Mechanism

NCEF guidelines describe the extent and mechanism of funding to the eligible projects. The main provisions are summarized as below:

- Projects sponsored by a Ministry/Department of the Government and submitted by individual/ consortium of organizations in the government/public sector/private sector are eligible for support in the form of loan or viability gap funding, as the IMG deems fit on case-to-case basis.
- Government assistance under the NCEF shall in no case exceed 40 % of the total project cost.
- The proposals by individuals/consortiums are to be submitted to the line ministry first, which, after due consideration shall bring them before the IMG.

- Projects that are being funded by any other arm of the Government of India or have *received grants from any other national/international body will be ineligible for* applying/funding under NCEF.
- In respect of time and cost overruns, a suitable accountability mechanism on lines similar to the one being followed in EFC/PIB projects/schemes shall be enforced strictly.

> The above mentioned funding mechanisms will not be able to realize the objectives at hand. A whole range of funding mechanisms will be required.

2.6 NCEF Activities: Based on Information in Public Domain

Information about the activities of the NCEF is not available in public domain (MoF 2012). A recent attempt by a Delhi-based organization in evaluating the Fund's performance under its present framework required filling a query, under the RTI Act, with the MoF. Information obtained through this route regarding the project proposals received and discussions in IMG meetings was available on this institution's Website (CBGA 2012).

According to a press report (ET 2012) based on the statement of the Union Finance Minister in the Parliament of India, the total tax revenue generated through "Clean Energy Cess" was Rs. 1,066.46 crore (actual) for the financial year 2010–2011 and Rs. 3,249.40 crore (revised estimates) for financial year 2011–2012. In respect of the current financial year (2012–2013), the budgetary estimates are of Rs. 3,864.20 crore. Till date, 15 projects envisaging total support of Rs. 1,974.16 crore out of the NCEF have been recommended by the IMG. During the financial year 2011–2012, the IMG recommended ten projects for NCEF support of Rs. 573.05 crore, while during the current financial year, as on date, five projects have been recommended for NCEF funding of Rs. 1,401.11 crore. This implies that more than 80 % of the corpus of NCEF is unutilized.

2.7 The Present Framework and Operation of NCEF: An Assessment

The launch of NCEF was welcomed because it raised the expectations that the Fund would help stimulate clean energy-related R&D, which also happens to figure high on the urgent needs of the clean energy sector in India. However, the allocations made from the Fund so far send out confusing signals. For instance, allocations for environmental pollution remediation projects and support to ongoing routine efforts on deployment of renewable energy (no disputes on the desirability and need for such initiatives) do not appear to be in line with the main objectives of the Fund.

Moreover, support for remediation of selected hazardous waste-contaminated sites supported by the Fund covers sites in industrial areas; whereas as per Para 2.1(iv) of the guidelines, "projects relating to environmental management particularly in the geographical areas surrounding the energy sector projects" is eligible. Given this, projects such as remediation of abandoned coal mining areas or affected areas near coal mines would be the obvious deserving candidates. Also, a number of approved projects under the NCEF constitute routine schemes and programmes of various ministries focused mainly on provision of clean energy for which technologies have been established and products identified. From the description available, there is no evidence of innovation in their delivery model either.

2.8 Key Findings from Review of Existing Structure and Operation of NCEF

A review of the NCEF shows that its present structure and framework for operation need to be sharpened and strengthened to improve its effectiveness and performance. The main points that emerge from review are as follows:

- The NCEF guidelines defining the eligibility of the projects for support are too broad based. This poses potential risk of diluting the focus of NCEF with adverse implications for research and innovation in clean energy sector in India. Especially so, in the absence of any identified targets and prioritization.
- The Fund lacks a vision, clearly defined targets, a road map to realize these targets, and a feedback mechanism to assess, learn, and improve.
- Innovative solutions (whether in technology, business models, and financial instruments) require a balance of actions along the innovation chain. Engaging with diverse stakeholders is critical in identifying such a balance in actions. Although the present framework provides for a mechanism to bring on board the experts and key stakeholders outside of Government systems, this opportunity has not been exploited.
- Funding limits and funding mechanism are not at all positioned to leveraging either domestic private investment or international resources and markets. Further, projects' ability to garner funding support from other sources should be rewarded and not penalized by making it ineligible for support from NCEF.
- The type and design of projects received for consideration, and the nature of discussion on them in IMG meetings point to an outlook that NCEF can be used freely to fund routine projects and schemes of various ministries as long as they meet a few general requirements. For instance, the discussions have largely focused on what revisions need to be made to a project proposal such that it fits better into the scheme rather than on the merits of the project in terms of its contribution in achieving the objectives of the Fund.

- There has been no mention, leave aside a structured discussion that the Fund needs to be proactive so as to encourage/invite projects which would promote research and innovation thus contributing to sustainable development of the clean energy sector.
- The requirement to apply through a Central government Ministry/department is faulty. Window for direct application should be there. Given the objectives of the Fund, a dedicated team/mission will be required to administer it. The present structure does not seem adequate and the most appropriate.

References

CBGA (2012) Centre for Budget and Governance Accountability. Available at http://www.cbgaindia.org

Central Board of Excise and Customs (2010) Notification on creation of national clean energy fund. http://pib.nic.in/newsite/erelease.aspx?relid=71517

Ministry of Finance (2011a) Office memorandum 18 Apr 2011. Available at http://finmin.nic.in/the_ministry/dept_expenditure/plan_finance2/Guidelines_proj_NCEF.pdf. Last accessed 6 May 2013

Ministry of Finance (2011b) Office memorandum 16 Jun 2011. Available at http://finmin.nic.in/the_ministry/dept_expenditure/plan_finance2/Format_Forwarding_NCEF_IMGC.pdf. Last accessed 6 May 2013

MoF (2012) Ministry of Finance response to the RTI filed by Gyana Ranjan Panda providing information on NCEF operations in FY 2011–2012

The Economic Times (2012) Available at http://articles.economictimes.indiatimes.com/2012-12-06/news/35647451_1_clean-energy-research-and-innovative-projects-financial-yearlast. Accessed 25 Feb 2013

Chapter 3
International Clean Energy Funds: A Review

Appropriate structure, and efficient administration and operation are critical for the success of clean energy funds irrespective of their objectives. This chapter examines the structure and operation of five international clean energy funds with a view to identifying the principles and practices which we may learn from improving the design and functioning of the NCEF. The following Funds have been examined:

a. Green municipal fund (GMF) of the federation of Canadian municipalities (FCM),[1] Canada.
b. California clean energy fund (CalCEF), California, USA.
c. Energy conservation promotion fund (ENCON), Thailand.
d. Clean energy finance corporation (CEFC), Australia.
e. Malaysian electricity supply industries trust account (MESITA), Malaysia.

3.1 Review of Funds and Learning for NCEF

This section presents stylized facts about each of the above fund. This is based on an analysis of the structure and operation of these Funds based on information available from the secondary sources. This is followed by a summary of key learning for NCEF presented in Figs. 3.1, 3.2, 3.3, 3.4, 3.5, 3.6, 3.7 and 3.8.

3.1.1 Green Municipal Fund

3.1.1.1 Key Design and Operation Features

Green Municipal Fund GMF (2012–2013), an endowment fund, was established in 2000 with the Government of Canada endowing FCM with CDN $125

---

[1] Federation of Canadian Municipalities (FCM) has been the national voice of municipal governments since 1901.

R. Pandey et al., *The National Clean Energy Fund of India*, SpringerBriefs in Energy, DOI: 10.1007/978-81-322-1964-4_3, © The Author(s) 2014

million—current endowment is CDN $550 million. The Fund was established as a long-term sustainable source of financing to provide low-interest loans and grants to support sustainable community development in Canadian municipalities that are more environmentally sustainable. GMF is more than a source of funding. It is a program that recognizes municipal leadership in sustainable development and works to help other municipal governments follow those examples through its capacity building and knowledge sharing programs. GMF funds can be used in combination with other funding. GMF is managed by FCM and operates at arm's length from the federal government.

FCM annually commits between CDN $65 million and CDN $90 million in low-interest loans and grants through GMF for initiatives that will significantly benefit the environment and are also likely to improve local economies and quality of life. Through GMF, FCM provides below-market loans and grants, as well as knowledge sharing and capacity building activities to support municipal initiatives that improve air, water and soil quality and protect the climate.

GMF is governed by FCM's National Board of Directors, which comprises over 70 elected municipal officials and affiliate members representing various geographic regions and various-sized communities throughout Canada. The FCM board is advised by a 15-member GMF Council. Five members of the Council represent and are appointed by the Government of Canada; the remaining ten are appointed by FCM. Of these, five represent municipal governments and the other five represent the non-profit and the private sector. GMF Council decisions are informed by a Peer Review Committee made up of 75 sector experts from across Canada. Each application to GMF undergoes an independent third-party technical assessment by two or three members of the Peer Review Committee. These assessments are then presented to GMF Council, which recommends a decision on eligible project proposals to the FCM National Board of Directors.

The GMF Council is guided by a council chair and two vice-chairs. One-third of Council members are Government of Canada representatives, one-third are elected municipal officials appointed by FCM Board of Directors, and one-third are external members representing the public, private, academic and environment sectors.

The GMF Council plays a key governance role for GMF and an advisory role for the FCM Board of Directors. The Council reviews completed peer review assessments on funding applications at its monthly meetings and makes funding recommendations to FCM's Executive Committee. Final decisions on GMF funding allocations are made by FCM's Executive Committee.

FCM offers GMF grants for Sustainable Community Plans, grants to conduct feasibility studies and field tests, loans and grants for Capital Projects reflecting the very best examples of municipal leadership in sustainable development— those that have high net environmental impact and that can be replicated in other communities. Applications are also assessed on the basis of project management, application quality, public engagement and municipal council or board of director's commitment. In keeping with FCM's goal to share lessons learnt from

GMF-funded initiatives with other communities, initiatives are also assessed on their innovation, potential for replication and potential for knowledge sharing. Applicants seeking funding for capital projects can submit applications at any time of the year. Applications received are evaluated and considered for approval within 4–5 months from the date they are received.

Since 2000, GMF funding has supported some of the most innovative, cutting-edge infrastructure projects in Canada, helping to drive sustainable community development and sustainable infrastructure practices across the country. Municipalities have built better transportation assets; constructed efficient and resilient buildings; diverted waste from landfills; made previously unusable land available for development; and improved soil and water quality. GMF-funded projects have saved millions of dollars for taxpayers and produced measurable results towards achieving Canada's sustainability goals.

While GMF is a relatively small program compared to federal and provincial/territorial infrastructure programs, it stands out among national sustainable development programs due in part to its strong integration of financial resources and capacity building services. Demand for GMF funding in recent years has consistently exceeded the available supply. To prudently manage this demand, FCM incorporated a more rigorous competitive approval process in 2012. This ensures that the most innovative capital projects with the most significant environmental outcomes receive funding, while enabling FCM to continue allocating funds equitably in all regions of Canada, in communities of all sizes.

3.1.1.2 Learnings for NCEF

- A multi-stakeholder 15-member advisory council and 75-member Peer Review Committee to advice and help and an independent third-party technical assessment of proposals are a striking feature of this Fund.
- GMF application process ensures transparency and accountability.[2] Application approval process takes 4 months.

---

[2] The Peer Review Committee, GMF Council, and the FCM National Board of Directors are integral to the application process. Each application undergoes an independent third-party technical assessment by two or three members of the 75-member Peer Review Committee. These assessments are presented to the GMF Council, which recommends a decision on eligible proposals to the FCM National Board of Directors. These recommendations are based on criteria outlined in the Agreement, including ensuring an appropriate balance between urban and rural communities as well as among regions within Canada. The FCM National Board of Directors ensures that due diligence is exercised in the decision process and makes the final decision on eligible project proposals.

- GMF capacity building program[3] has developed a number of successful initiatives to transfer knowledge and build capacity across the country.
- By strategically allocating funds to the best projects and studies, and sharing the lessons and expertise from those initiatives with other municipalities across Canada, effectiveness of GMF increases manifold.[4]
- GMF funds can be used in combination with other funding.

3.1.2 The California Clean Energy Fund

3.1.2.1 Key Design and Operation Features

Founded in 2004, the California Clean Energy Fund (CalCEF 2004) is a non-profit organization working to accelerate the movement of clean energy technologies along the continuum from innovation to infrastructure using tools from finance, public policy and technological innovation. CalCEF was created with $30 million grant from shareholders to create a de novo organization to stimulate clean energy technology development for California. This was set as a condition of Pacific Gas and Electric Company (PG&E) bankruptcy reorganization at the direction of the California Public Utilities Commission.

The private companies in which the "Fund" invests create technologies, design products and provide services. CalCEF uses three platforms to run its programs in Clean Energy namely the following:

- CalCEF Capital for their investment programs and
- CalCEF Innovations, which is a centre for market strategy, policy and product development.
- CalCEF Catalyst, an industry acceleration platform.

---

[3] GMF's capacity building program includes the following: GMF Webinar Series: interactive, Web-based workshops that feature presentations from sector experts and a GMF-funded municipal practitioner; FCM Sustainable Communities Conference: biennial national conference; Partners for Climate Protection, which aims to mitigate climate change through reduced greenhouse gas emissions; FCM-CH2M HILL Sustainable Community Awards: Held annually, the program recognizes municipal leadership in sustainable community development; and FCM Sustainable Communities Mission: GMF organizes study tours that enable elected and senior municipal staff officials from across the country to visit and learn about leading sustainable community development sites and protects, some of which are GMF-funded. In addition to the above initiatives, GMF develops case studies of funded projects, and shares them through a searchable database available to the public online (see the GMF section of FCM's website (www.fcm. ca/gmf). GMF also organizes capacity building workshops on issues related to its funded sectors, and develops tools and resources for municipalities (also available online).

[4] Three integrated, collaborative functions support this goal: research; capacity building; and communications. FCM conducts research related to GMF funding sectors, including identifying key results and lessons learned from GMF-funded initiatives. It builds the capacity of municipal governments to implement sustainable community development projects and practices through tools and training. Finally, FCM transfers knowledge and performs communications activities related to outreach and promotion, publications and web development, and media relations activities.

CalCEF pioneered first-venture capital fund of the clean energy industry viz the CalCEF Clean Energy Angel Fund. The CalCEF Clean Energy Angel Fund is the first seed stage-focused investment vehicle in the clean energy market. The Fund is designed to address a persistent and increasingly significant problem in the clean energy industry—the absence of funding for companies at the earliest stage of their development.

Investment programs of the CalCEF Capital include the following:

- CalCEF Ventures: Established to create institutions and investment vehicles that accelerate the adoption of clean energy technologies. It makes for-profit investments into new funds supporting the public interest in advancing clean energy, catalyzing further private capital and recycling its profits via an "evergreen" investment strategy into further fund creation. Ventures also make strategic grants to support the launch and growth of important institutions advancing the broad CalCEF agenda. CalCEF's investment strategy focuses exclusively on clean energy, including renewables, energy efficiency, energy storage and enabling technologies and services. Under the terms of the agreements, the venture capital firms will make equity investments in clean energy companies on behalf of CalCEF.[5] By working with highly qualified investment firms, each with their own investment expertise, CalCEF takes a blended approach to the market that mitigates risk and maximizes returns. The goal is to support a wide range of opportunities, including both later-stage and early-stage opportunities, where CalCEF funding will make a difference.
- CalCEF Clean Energy Angel Fund[6]: It is the first seed stage-focused investment vehicle in the clean energy market. The Fund is designed to address a persistent and increasingly significant problem in the clean energy industry—the absence of funding for companies at the earliest stage of their development, the so-called valley of death problem. The Angel Fund is a separate, for-profit entity independent of CalCEF with multiple individual and institutional LPs. Its objective is to produce an attractive financial return on investment and to meet a critical need in the emerging clean energy industry.
- CalCEF Innovations: It was formed in 2008 to design and pilot business models, financial products and public policies that grow clean energy markets and accelerate adoption of clean energy technologies. These goals are achieved primarily through "Entrepreneurs-In-Residence" Program.

---

[5] CalCEF had allocated $8.5 million to each of the three funds for a total of $25.5 million. Nth Power and Draper Fisher Jurvetson (DFJ) each directly manages an investment portfolio totalling $8.5 million, with DFJs allocation to be managed through DFJ AltaTerra, a DFJ affiliate fund launched to make investments in the clean technology sector. These managers will also match each dollar invested on behalf of CalCEF with its own investments in order to maximize market impact. CalCEF also participated as a limited partner in VantagePoint Venture Partners. The remaining $4.5 million had been reserved by the CalCEF Board for future program development.

[6] CalCEF Clean Energy Angel Fund is a seed/start-up stage investment fund in the clean energy and related technologies market, including energy efficiency, renewable energy, power reliability and alternative energy.

3.1.2.2 Learnings for NCEF

- CalCEF's investment strategy focuses on identifying and solving gaps and barriers that are slowing expansion of clean energy markets and adoption of clean technologies.
- CalCEF acts as a leader, organizer and investor in addressing critical barriers in clean energy industry in the USA.
- CalCEF's diverse stakeholders—leading investment firms, policy makers, academics, scientists and advocates—provide a constant stream of insights into the challenges facing this unique and critical industry.

3.1.3 Energy Conservation Promotion Fund, Thailand

3.1.3.1 Key Design and Operation Features

The ENCON Fund (UNDP 2012) is an extra-budgetary fund established in 1992 to provide financial support for implementation of the ENCON Act for promoting energy conservation in Thailand. ENCON fund provides working capital, grants and subsidies for investment in energy conservation programs in both public and private sectors.

The Ministry of Energy manages the Fund through ENCON Fund Committee[7] with the guidance of the National Energy Policy Council (NEPC). Of the total budget (THB 7 billion annually), around two-thirds is managed by the energy policy and planning office (EPPO), while department of alternative energy development and efficiency (DEDE) is responsible for managing the remaining one-third.

The EPPO provides grants to government agencies, universities and NGOs for various projects, besides implementing a demand-side management (DSM) bidding program to encourage business operators to invest in higher-energy-efficiency machines/equipment. DEDE also implements a wide range of financial mechanism, such as the Thailand Energy Efficiency Revolving Fund, the ESCO venture capital and tax incentives to promote energy conservation and increase the share of renewable energy in the total energy mix in the country.

---

[7] ENCON Fund Committee is chaired by the Deputy Prime Minister and EPPO serves as the secretariat of the Committee. The mandate of the Committee includes: (i) to propose energy conservation promotion policies, goals and measures to NEPC; (ii) to propose to NEPC guidelines, criteria, conditions, and priorities for the disbursement from the ENCON fund; (iii) to prescribe regulations on the criteria and procedures for applications, grant allocations or subsidies from the ENCON fund; (iv) to allocate appropriations from the ENCON fund; and (v) to propose to NEPC contribution rates to be imposed on petroleum products for the ENCON fund.

In the case of the ENCON fund, specific objectives and quantitative targets
to be achieved within a period of time are set in the ENCON program. The
Five-Year Energy Conservation Program has been developed to provide a guideline for the utilization of the ENCON fund. Phase 3 of the ENCON program
(2008–2011), for instance, aimed to increase energy efficiency by 10.8 % from
business as usual and increase the share of renewable energy development to
15.5 % of the total energy consumption. The specific targets set by the ENCON
program ensure that the ENCON fund is managed strategically according to the
government priorities on renewable energy and energy efficiency. Setting specific policy objectives and targets enables the monitoring and evaluation of the
performance of the fund. The effectiveness of the fund can then be monitored
and evaluated based on the performance since there is a clear link between the
amount of resources spent with the achievement of policy targets and performance indicators.

3.1.3.2 Learnings for NCEF

- Specific objectives and quantitative targets along with a time frame are set in
  the ENCON program. For instance, a Five-Year Energy Conservation Program
  has been developed to provide a guideline for the utilization of the ENCON
  fund.
- To implement the Revolving Fund, the DEDE has collaborated with commercial
  banks. The ESCO Fund is being managed by the professional fund managers.
  The fund managers proactively work with the main target group, SMEs, as a
  single-window facility. This increases the overall efficiency of the program.
- Periodic assessment of the effectiveness of programs is a key feature of this
  fund. For instance, following a review of the Revolving Fund conducted by
  DEDE, it was found that the main beneficiary of the program was large enterprises. Based on this review, a discussion was started within DEDE where
  a specific program targeting SMEs was considered necessary. In 2008, the
  ESCO fund was introduced specifically for targeting SMEs. The procedures
  and criteria of the ESCO fund were developed considering the characteristics
  of the SMEs. As a result, non-profit organizations were appointed as the fund
  managers.

3.1.4 The Clean Energy Finance Corporation

3.1.4.1 Key Design and Operation Features

The CEFC (ACF 2011) was announced under the Clean Energy Future Package—
the Australian government's package to put a price on carbon pollution in 2011.
The CEFC is a mechanism to help mobilize investment in renewable energy,

low-emission and energy efficiency projects and technologies in Australia to address the barriers currently inhibiting investment. The CEFC will start investing from July 2013. The CEFC is an independent institution, established under the legislation and removed from annual budget cycles and politics.

An amount of $10 billion will be seeded into the CEFC over 5 years for investing in deployment and commercialization of emerging renewable energy technologies such as solar PV, solar thermal and geothermal (CEFC 2012). Funding will be provided in two streams: 50 % of funding is reserved for renewable energy projects ($5 billion). The other 50 % of funding is available for renewables plus more general clean energy projects—energy efficiency, low-emission technologies and building manufacturing businesses to underpin these sectors ($5 billion). The CEFC will not provide grants as in other government programs, but rather will invest with private investors using loans, loan guarantees and equity. In this way, projects that would otherwise not be funded will attract private investment to get off the ground. The CEFC will be commercially oriented, staffed by experienced investment, banking and clean energy experts. With the $10 billion of public money, the CEFC is expected to leverage up to an additional $100 billion of private investment in the coming decades.

CEFC as a publicly owned and accountable entity has to operate in a manner consistent with the Government's expectations and within its investment mandate. The enabling legislation by the Government sets the framework for CEFC and to determine how the corporation is directed, controlled and held to account. The investment mandate also includes guidance to the Board on the objectives and parameters, under which the CEFC will operate, with Board having the overall responsibility for investments. According to the recommendations, the board should comprise people with skills and experience in banking and finance, investment management, venture capital and private equity, clean energy sector technologies and engineering, and/or the environmental sectors.

3.1.4.2 Learnings for NCEF

- The CEFC Board will comprise people with skills and experience in banking and finance; investment management; venture capital and private equity; clean energy sector technologies and engineering; and/or the environmental sector.
- The CEFC will build on existing government grant funding for R&D [that will continue to be delivered through the recently announced Australian Renewable Energy Agency (ARENA)] and thereby plug the gap between R&D and commercialization.
- Fund allocation: 50 % or more of funds will be allocated to the renewable energy, and up to 50 % will be allocated to the low emissions and energy efficiency.

3.1.5 Malaysian Electricity Supply Industries Trust Account

3.1.5.1 Key Design and Operation Features

MESITA was officially launched in July 1997 (MESITA 1997). The contributors to the fund are the power-generating companies, i.e. TNB Generation Sdn. Bhd. and independent power producers (IPPs) in Peninsular Malaysia comprising Genting Sanyen Power Sdn. Bhd., Port Dickson Power Bhd., PowertekBhd., Segari Energy Venture Sdn. and YTL Power Generation Sdn. Bhd. Their contribution is voluntary, and they contribute 1 % of their electricity sale (total annual audited turnover) to the Peninsular Grid or the transmission network.

The Electricity Supply Industries Trust Account Committee manages the trust account. The Committee comprises of representatives from The Economic Planning Unit, Prime Minister's Department, Ministry of Energy Green Technology and Water, Energy Commission, Ministry of Finance, Ministry of Rural Development, Tenaga Nasional Berhad (TNB), and Six electricity generating companies. The Committee is chaired by the Secretary General, Ministry of Energy Green Technology and Water. A Technical Committee assists the Electricity Supply Industries Trust Account Committee in evaluating applications for funding from the trust account. MESITA exclusively targets the Electricity Supply in various sectors. MESITA is used in supporting Rural Electrification Program, R&D Programs, new renewable sources of energy projects, human resource development programs for the industry, energy efficiency projects, and development and promotion of the electricity supply industry.

3.1.5.2 Learnings for NCEF

- A Technical Committee assists the Electricity Supply Industries Trust Account Committee in evaluating applications for funding from the trust account.
- The guidelines of MESITA clearly identify the specific projects which can be considered by MESITA, expected output from project, expected organizational outcome, and expected sectoral and national impacts of the projects.
- Financial contribution and active participation of utilities is an interesting feature of the fund.

3.2 Key Lessons from Review of International Clean Energy Funds

Key lessons on various aspects of successful international clean energy funds are summarized in the Figs. 3.1, 3.2, 3.3, 3.4, 3.5, 3.6 and 3.7.

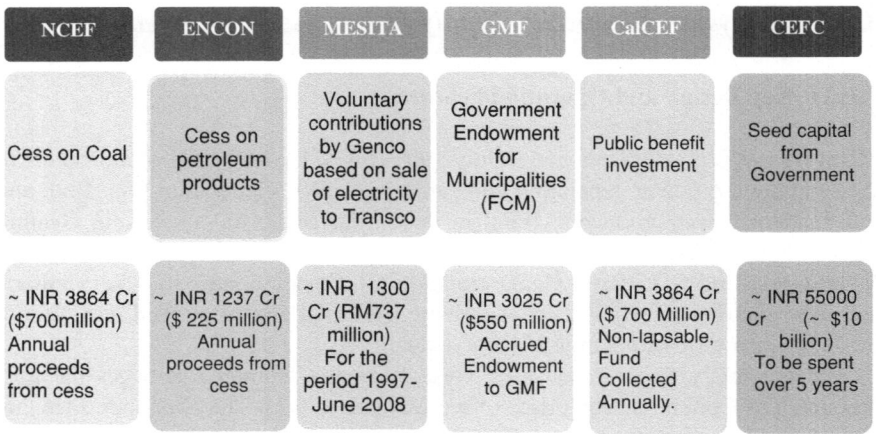

Fig. 3.1 Source of funding and size of funds—NCEF versus international funds

NCEF	ENCON	MESITA	GMF	CalCEF	CEFC
To fund research and innovative projects in clean energy technologies.	To provide financial support to the implementation of the ENCON Act and providing assistance for the energy conservation in Thailand.	To provide financial assistance to the electricity supply projects, R&D and energy efficiency.	Providing funding and knowledge to municipal governments and their partners for municipal environmental projects.	To accelerate the movement of clean energy technologies along the continnum from innovation to infrastructure using tools from finance, public policy and technological innovation.	To help mobilize investment in renewable energy, low emissions and energy efficiency projects and technologies in Australia thus addressing the barriers currently inhibiting investment.

Fig. 3.2 Objectives of the funds—NCEF versus international funds

NCEF	ENCON	MESITA	GMF	CalCEF	CEFC
Ministry/Department of the government; and Individual/consortium of organizations in the Government/public sector/private sector.	Government agencies, NGO's, Universities, SME's, Public/Private sector entities engaged in projects promoting energy conservation.	Private/public agencies supplying electricity and engaged in related fields, Universities.	All municipal governments and their partners.	Private Producers and entrepreneurs.	Private Sector.

Fig. 3.3 Target beneficiaries—NCEF versus international funds

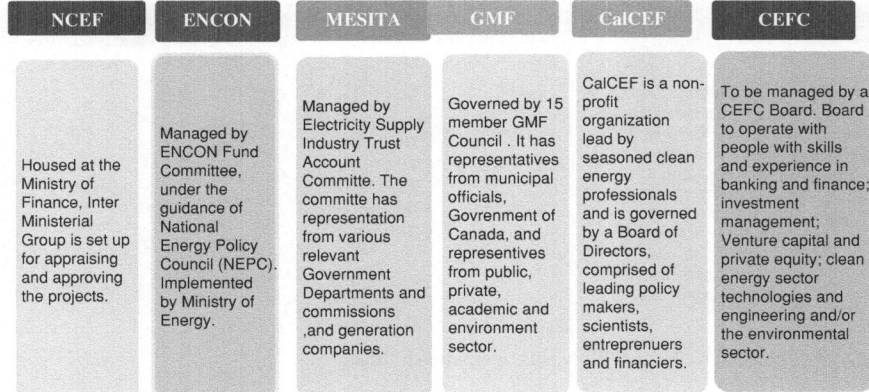

Fig. 3.4 Administration and management—NCEF versus international funds

Fig. 3.5 Fund allocation—NCEF versus international funds

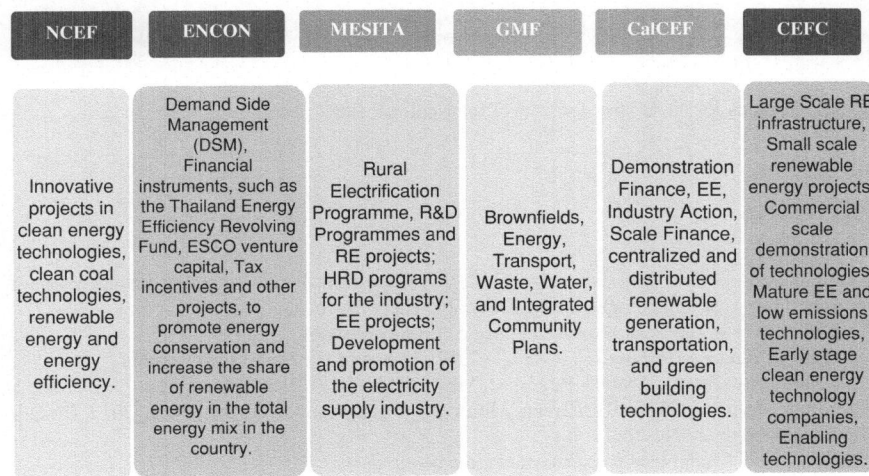

Fig. 3.6 Focus areas for support—NCEF versus international funds

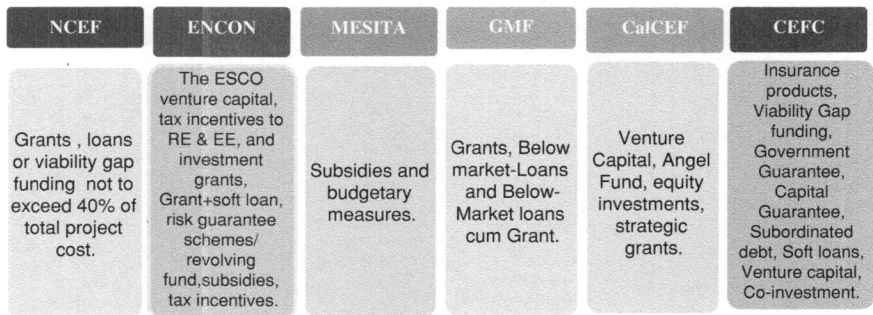

Fig. 3.7   Financial tools—NCEF versus international funds

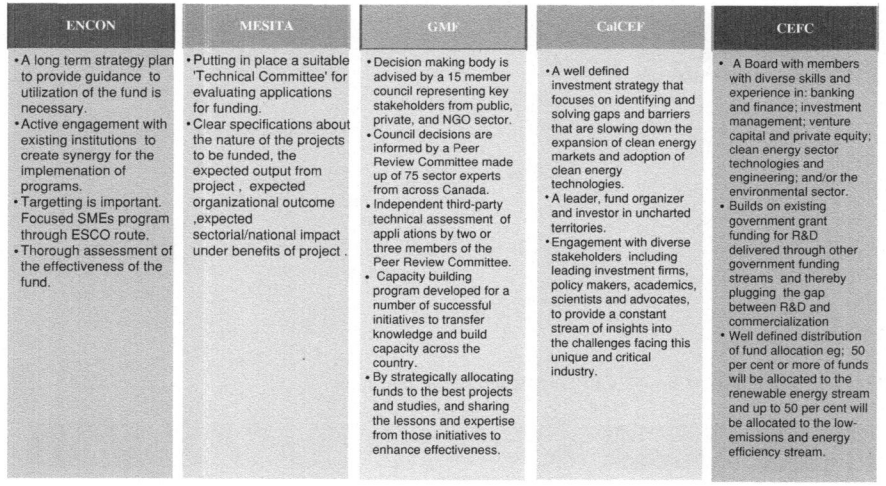

Fig. 3.8   Lessons for NCEF from review of international funds

References

Australian Conservation Foundation (ACF) (2011) Report on the clean energy finance corporation: helping Australia compete in the renewable energy race

CALCEF (2004) Available at http://calcef.org/catalyst/programs/

Clean Energy Finance Corporation (CEFC) (2012) Report of the expert review panel

Green Municipal Fund—Annual Report (GMF) (2012–2013) Available at http://www.fcm.ca/Documents/reports/GMF/2013/Green_Municipal_Fund_Annual_Report_2012_2013_EN.pdf. Last accessed on 20 Nov 2013, p 1

MESITA (1997) Available at http://mesita.kettha.gov.my/

UNDP (2012) Environment and energy program's case study report on Thailand energy conservation fund

Chapter 4
NCEF: Aligning Activities with the Objectives

Analysis in Chap. 2 brings out that the actual disbursements/approvals so far from NCEF are aligned more with ongoing routine programmes/missions of various ministries/departments than with the stated objectives of the fund. Also, utilization of funds from NCEF has been slow.

However, this is an expected outcome in the absence of a well-thought-out framework for allocation of funds. In this context, the following framework is proposed.

4.1 Niche for the Fund and Value Addition of the Fund Needs to be Spelt Out Clearly so that it is Properly Understood by the Stakeholders

As stated earlier, the objectives of the fund as per NCEF guidelines are as follows: "Funding research and innovative projects in clean energy technologies. Any project/scheme relating to innovative methods to adopt to clean energy technology and research and development shall be eligible for funding under the NCEF".

It is important to note that the above statements do not distinguish between

(a) Encouraging the development of innovative clean energy technologies per se (through R&D in innovation and demonstration stages)
(b) Supporting innovative methods of adopting clean energy technologies (i.e. targeted deployment and untargeted diffusion)

Hence, the fund can either support both types of initiatives sequentially or choose to support one or the other. Starting with technology development through application-oriented R&D, the fund can in the medium to long term support activities in

R. Pandey et al., *The National Clean Energy Fund of India*, SpringerBriefs in Energy,
DOI: 10.1007/978-81-322-1964-4_4, © The Author(s) 2014

the nature of barrier removal and market penetration for large-scale deployment. Alternatively, it could support an initiative focussing on R&D come pilot demonstration with relatively less focus on innovative deployment and market penetration projects. A number of factors will determine the sequencing and relative weights of these activities. A further discussion on these is provided later in this chapter.

> NCEF should support both innovation in clean energy technology as well as ways of adoption/deployment of clean energy technologies that may have been piloted but await innovative application for supporting market creation and deployment.

The former will refer to technology innovations that may be at the cutting edge of research and need R&D support from NCEF to ready them to be piloted. The innovative application methods on the other hand refer to new and creative ways of upscaling existing/piloted technologies and handling market penetration barriers that increase their adoption and large-scale deployment.

It is important to note here that a clean energy fund with mandates as that of NCEF is often designed to fund options where benefits accrue over the long term. For, programmes such as technology research, development, and demonstration programmes require a longer time frame (5, 10 or more years) than is typically allowed by other approaches/policies such as renewable portfolio standards which are generally aimed at jump-starting markets for commercially ready technologies.

With this background, the NCEF may consider the following as its core constituencies:

- Acting as a catalyst to help boost development of a robust clean energy industry
  - Identifying technology and innovation needs and establishing a development plan for the same. Within this, a strategy for prioritizing.
  - Financial and institutional support for accelerating clean energy technologies and innovative projects.
  - Identifying skill development needs and developing a skill development plan.
  - Knowledge creation and sharing.
- Acting as an anchor for establishing linkages and cooperation with international institutions/programmes in areas of core mandate of NCEF
- Acting as an anchor for synergy between other government efforts in areas of core mandate of NCEF

And in the larger context

- A dedicated NCEF team with appropriate expertise and accountability will be necessary to achieve the above.

4.2 Acting as a Catalyst to Help Boost Development of a Robust Clean Energy Industry

4.2.1 Identifying Technology and Innovation Needs and Instituting a Development Plan for Same

R&D in the energy sector is critical to augment and diversify our energy resources and to promote energy efficiency. R&D requires sustained and continued support over a long period of time. In India, R&D has not been allotted the resources that it needs and this is especially true in the case of energy-related R&D. There is a need to both substantially augment the resources made available for energy-related R&D and allocate these strategically according to its needs and priorities, specific circumstances and capacities, and specific framework conditions.

The first critical priority in this context would be a suitable energy technology policy and an assessment of technology needs. The technology road mapping can add substantial value to technology policy. Technology road mapping is one of many technology planning tools that countries undertake when identifying, selecting and investing in technologies that are needed to meet their needs. Industry, scientific and technical research institutions, public sector institutions such as DST, MNRE, Strategic Knowledge Mission, National Innovation Foundation, NGOs and consumer groups will be important stakeholders and collaborators in this exercise. This exercise will have to be based on a dynamic strategic vision which is frequently updated. Next, from the menu of clean technologies available, India needs to choose those that it can tweak to suit its needs within its constraints.

Having identified the technology needs, the next step would be a mapping of various ongoing efforts, both institutional and others. This will help NCEF in determining its role from other existing programmes, thus checking overlaps and maintaining focus of different initiatives/programmes, thereby enhancing the overall effectiveness of various initiatives. The gaps so identified will guide the fund's clean energy technology and innovation programme. An outline of this process is given in Fig. 4.1.

In India, progress in clean energy sectors, in terms of availability of credible resource assessment data; stages of technology development; enabling fiscal and regulatory policies; and end-user awareness and acceptability has been at different levels. Owing to this, the following assume importance in designing a clean energy technology and innovation programme:

- There is a lot of scope in existing technology, be it efficiency improvement or cost reductions. A balance will need to be found between promoting further innovation in existing technologies and next-generation R&D.
- There is room for debate on how NCEF should strike a balance between supporting old model of R&D and innovation models based on trial-and-error methods which have more scope for involving grass root ideas and young researchers.

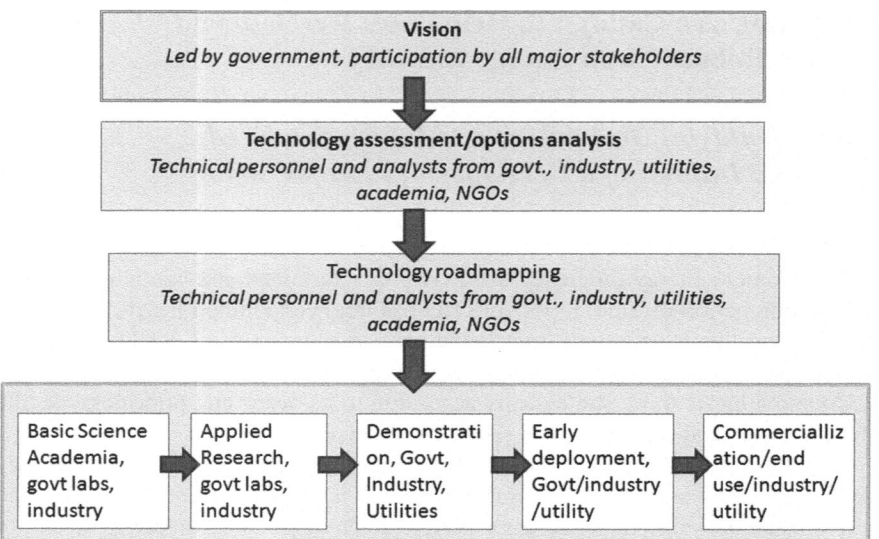

Fig. 4.1 Identifying technology needs. *Source* Chikkatur and Sagar (2007)

4.2.1.1 Specific Energy Sub-sectors and Prioritizing

In this context, an important question is that within the domain of research and innovation in clean energy technologies and harnessing renewable energy, what guiding principles should be used in identifying the specific energy sub-sectors and a priority list within that? We suggest the following framework:

- inclusive development and energy security to all;
- meeting the commercial energy needs of the unserved population and in providing community-based local solutions;
- research and development of key sectors and technologies;
- building a robust clean energy industry that becomes an important driver of economic strength.

With the broad vision of IEP and other considerations as above, the coverage of NCEF, thus, could span the spectrum of both supply-side and demand-side issues. Of critical importance is research and analysis for the energy policy to outline technology road maps. The NCEF should encourage and fund such studies in a number of institutions on a long-term basis and should also commission studies to independent experts and consultants. A number of academic institutions should be developed as centres of excellence in energy research. Besides, coordinated research in all stages of innovation chain should be supported.

In this context, and 'energy policy, technology and innovation forum' may be set up which can serve as a platform for recognizing and rewarding innovation, and sharing knowledge and best practices. The bigger ambition would be that the important results/best practices feed into political process and international discussions.

In what follows, an attempt is made to identify potential areas in identified energy sub-sectors for support/intervention by NCEF.

4.2.1.2 Coal

Energy demand growth in the next 20 years is projected to be staggering. The Central Electricity Authority (CEA) has estimated that meeting our electricity demand by 2017 will require total installed capacity of 280 GW of which 80 GW of new capacity is expected to be based on coal.[1] Given the massive predicted growth rates for coal, it is necessary to focus on ways of providing coal-based energy (power, light, heat and mobility) with a reduced level of resource, environmental pollution and climate impact.

The coal cess is not a carbon tax, and it does not establish a price of carbon. It is a cess on coal producers and importers and thus has implications for price of coal-based energy. It may do little to encourage clean coal technologies, processes and methods in the entire supply chain unless some of the revenues are directed towards encouraging these activities.

There have been some efforts in this direction,[2] yet there exists considerable scope of enhancing the resource conservation and minimizing environmental impact around the coal value chain.

Opportunities for NCEF in Coal Sector

- R&D in coal mining is minimal. Development and adaptation of technologies for mining low-ash coal and efficient coal handling is required. Also, In-pit crushing and conveying technology using mobile/semi-mobile crushers can form an alternative to diesel-fuelled dumper transport where large volumes of coal and overburden need to be handled. This system has significant

---

[1] Coal demand for power sector is projected to increase from 308 Mt in FY 2007 to 750 Mt in FY 2017 registering a CAGR of 9.3 % for the period.

[2] For promotion of clean coal technologies, action has been initiated with the creation of Indo–US Working Group and Asia–Pacific Partnership. Twelfth plan has specified targets. Under a GEF, UNDP and MoEF funded project a CBM recovery and commercial utilization project was approved with the objective of harnessing methane to minimize safety risks, to mitigate environmental impact and to utilize potential energy source.

environmental benefits in terms of reducing airborne dust and carbon emissions. Coal-bed methane and underground coal gasification are other areas which would need support with technology adaptation.

- Innovative ideas on general environment management around the coal mines, coal washeries, rail sidings and other coal-utilizing plants, especially those ideas that provide synergy to the ongoing initiatives of environmental management programmes and CSR programmes including those on ash management.
- Coal beneficiation reduces the ash content in the coal and improves its thermal efficiency and reduces operation and transport costs of power plants and other users. A study by Chelliah et al. (2007) recommended the "levy of an eco-cess differentiated on the quality of coal to provide suitable incentives for coal beneficiation. It also suggested that the cess should be supplemented with reforms in the power sector. Revenues generated from cess may be used to set up a clean coal fund which could be utilized for setting up infrastructure for coal washing, selective mining and related research and development". National policies should evolve to enhance support for coal washeries. Innovative ideas of improving efficiencies in coal washing include integration of fine coal circuits in washeries. Research and development of technologies use less water for coal washing and for efficient utilization of washery rejects for energy generation such as fluidized bed technology (FBC). For example, a low-water-utilizing washing technology could be the dry beneficiation system using radiometric techniques or dry beneficiation of coal by All-air Jig or Variable Wave Jig Technology which can be applied in the existing arrangements with small modifications.
- The adoption and success of supercritical technology will depend largely on the coal quality and its assured supply. The experience in manufacturing, supply and operations of high-pressure and high-temperature main plant equipment is limited in the country. The long-term impact of high-pressure and high-temperature profiles on the boiler and related components' life is not yet known, and closer collaboration between technology suppliers and generators will be important initially (Fig. 4.2).

4.2.1.3 Renewable Energy

Renewable energy in Indian context can support and serve a number of developmental objectives. For example, renewable energy has the potential to provide a buffer against the energy security concerns of our country; it offers a hedge against fossil fuel price hikes and volatility; and off-grid renewable energy can meet demand in unserved remote rural areas while addressing India's poverty eradication and job creation goals.

Market assessments indicate that India could eventually be the largest renewable market in the world given the abundance of renewable energy resources.[3] At the sectoral level, small hydropower (SHP) and wind energy are relatively mature with significant local capacity, although there are opportunities in the manufacturing of

---

[3] US Department of Commerce and International Trade Administration (2008).

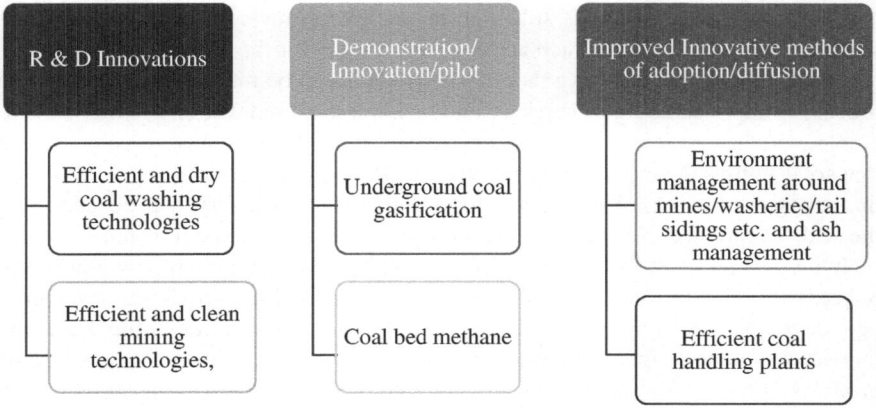

Fig. 4.2 Prioritization options for cleaner coal technologies

products and equipment and demonstration of technology and project development in these sectors. Contribution of waste-to-energy and solar energy is very small, while electricity generation from solar thermal, geothermal and ocean power is non-existent. Geothermal and tidal energy sectors offer good scope for R&D. This is an indicator of the opportunity that is available in harnessing the full potential of these sectors.

Renewable energy technology deployment when based on proper resource assessment has the potential to provide energy security and economic development in urban, rural and select industrial set-ups. In the rural set-ups, the example of such projects could be innovative ideas that integrate renewable energy within larger supply chains within rural economies including those based on agriculture, forestry, traditional manufacturing and green tourism.

Expanding the installed capacity of wind, solar and biomass technologies is crucial. The scale at which renewables could be deployed relies to a great extent on their commercial competitiveness, which in turn depends heavily on the success of technology development and diffusion. An attempt is made to identify opportunities for NCEF in identified renewable energy sub-sectors as follows:

4.2.1.4 Solar Energy

The solar resource in India is distributed evenly over a larger geographical area; therefore, it can present for a greater opportunity for reaching out to unserved areas, especially those where conventional modes of energy systems face a constraint either due to remoteness of location or due to any other factors.

Government of India has launched several initiatives including flagship project on deployment of solar technologies with ambitious installed capacity targets. The project encourages deployment of solar photovoltaic technologies and solar thermal technologies for grid-connected, off-grid, heating, drying and cooling applications.

The cost of solar power is still high in absolute terms compared to other conventional sources of power such as coal. The need is for the conductive conditions that drive down the costs towards grid parity. This can be made possible through a process of rapid scale-up of capacity and technological innovations. While considerable evidence exists to show that costs have come down in past three years and that solar will achieve grid parity by 2017–2018 and coal parity by 2025, however, this recognition is based on the assumption that cost trajectory will depend upon the scale of global deployment and technology development and transfer.

India's target of 20 GW of installed solar capacity by 2022 is highly ambitious. Success in meeting this target will require international collaboration in technology development, support for development of a local manufacturing base and innovative financial mechanisms to enhance its commerciality. In spite of some progress, solar energy sector is faced with a number of barriers in the supply chain including the sustainable delivery models (MNRE 2011).

Barriers

- Solar technologies are at a nascent stage in India, and there are considerable risks in execution of projects.
- Crystalline cells and modules are comparatively easier to execute and less risky as manufacturers generally guarantee the products for 20+ years. However, newer technologies such as thin film and concentrated PV, though they have lower upfront costs, are unproven and therefore considered more risky.
- The returns of a solar project are highly sensitive to radiation levels. High-quality solar radiation is a prerequisite for proper market assessment and project development. Hence, solar radiation assessment is a very important activity and typically requires several months for ground measurement of solar radiations. Any error in solar resource estimation adds an uncertainty to expected future returns. As of now, on ground, solar radiation data are sketchy and the simulation models are at preliminary stage.
- Evacuation of the electricity generated from large-scale power plants located in isolated areas is a potential challenge. It may require development of new transmission lines which are often controversial, both because of their expense and because of the potential for damaging property and environment.

Opportunities for NCEF in Solar Energy

- Technology refinements in decentralized solar energy systems: These include solar water heating systems, home lighting systems which include solar lanterns, solar cooking systems, solar pumps and small power-generating systems.
- In the PV sector, there is demand for thin-film solar cell technology, technology for megawatt-scale power generation and improvements in crystalline silicon solar cell/module technology. Building integration for PV and solar thermal systems is also an area of opportunity.

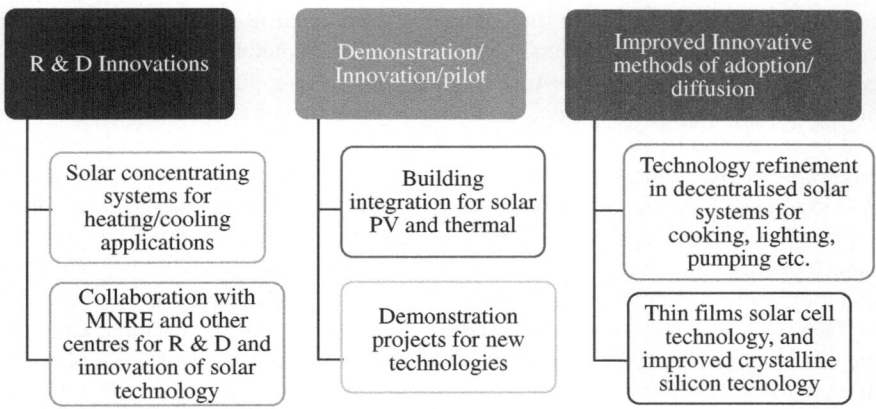

Fig. 4.3 Prioritization options for solar energy technologies

- Demonstration projects for new technology.
- Support for R&D for development in various solar concentrating systems for heating/cooling applications. It is reported that knowledge base exists in engineering colleges and research institutes for carrying forward innovation in this area. For instance, an engineering student is reported to have developed a technology for reducing the cost of solar thermal heating by designing an indigenous low-cost solar reflector (Financial Express 2013).
- To collaborate with MNRE and other institutions for technology development and adaptation, manpower development, innovative product delivery and service models and covering performance uncertainties and risks of new technologies (Fig. 4.3).

4.2.1.5 Wind Energy

As of March 2012, renewable energy accounted for 12.2 % of total installed capacity, up from 2 % in 1995. Wind power accounts for about 70 % of this installed capacity. By the end of November 2012, wind power installation in India had reached 18.3 GW (MNRE 2012). The total capacity potential is estimated to be 49,130 MW.

India's robust domestic market has transformed the Indian wind industry into a significant global player. The success on the Indian wind market can be attributed to the quality of the wind resource, to domestic tax incentives and, to a lesser extent, revenue from the Clean Development Mechanism.

Upcoming technological developments include wind forecasting to enable integrated grid management and more efficient generation. The MNRE and CERC recently commissioned PGCIL to study and identify transmission infrastructure for renewable energy capacity addition during the 12th plan period. After the extensive consultation with stakeholders including the state nodal agencies, the final report called "Green Energy Corridors" was released in September 2012. It discusses issues of intra- and inter-state transmission system strengthening and augmentation, establishment of a Renewable Energy Management Centre, improved forecasting to address variability aspects as well as grid integration issue of large-scale renewable energy generation. An investment of approximately INR 42,257 crores is being planned for the development of this corridor by 2017. Out of this amount, approximately INR 20,466 crores is likely to be invested in strengthening the inter-state transmission system. This initiative if implemented successfully could be a major driver for the development of renewable energy sector in India.

There is also a rising interest in offshore wind developments in India, although there has not yet been any significant progress. The trend of recent installations is moving towards better aerodynamic design; use of lighter and larger blades; higher towers; gear and gearless machines; and variable-speed operation including using advanced power electronics. The machines with permanent magnet generators that are suitable for moderate wind regime are also being installed in the country.

Barriers

- Efficiency Issues: Wind energy forms about 10 % of the total installed capacity of the country but contributes less than 3 % to the country's power generation. Indian wind farms operate at 15 % of its total capacity.
- Environmental Issues: Concentration of wind turbines in an area increases the average temperatures.
- Transmission Constraints: The transmission network building measures by the wind energy-rich states remain extremely crucial for the sustained development of the sector. Creating power evacuation infrastructure for renewable is challenging due to
  - Remote location of renewable energy potential.
  - Intermittency of renewable energy (e.g. fluctuating supply from renewable energy sources creates a complexity in the grid). One option is pooling of geographically disperse intermittent sources so that average power at pooling station does not have more fluctuation.
  - High costs of transmission infrastructure.

Opportunities for NCEF in Wind Energy

- Introducing innovative channels of deployment (large/small scale) that result in economies of scale and hence reduced project costs (e.g. deployment of small-scale wind farms that can be integrated with solar and biomass using micro-grid).
- Innovative deployment strategy that complements the outreach of clean wind energy to local and rural farmers, thereby improving their economic well-being through ways such as adding more value to the human labour and farm produce, thereby stabilizing the income generation of such populace.

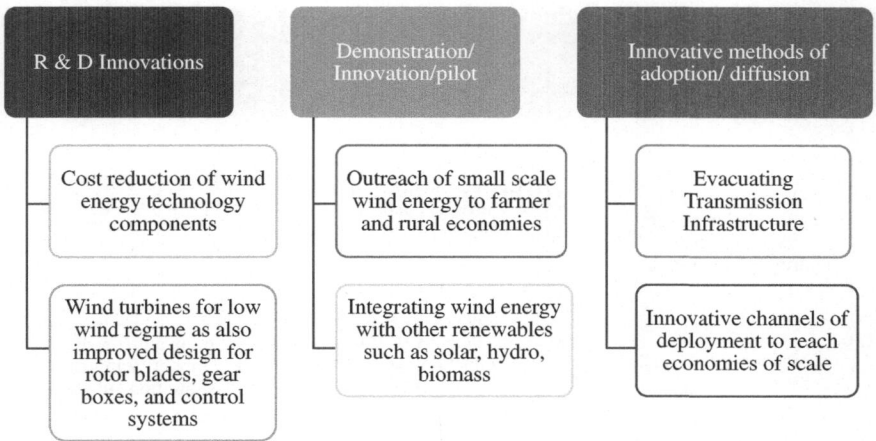

Fig. 4.4 Prioritization options for wind energy technologies

- Research and development on cost reduction of components used in wind energy technology. A possible approach could be incubating technologies that have high future potential with active participation of national institutions.
- Integrating different renewable sources such as solar and wind, which produce peak energy during different times of the day. This will reduce supply fluctuation and lead to better utilization of transmission system.
- Judicious planning of transmission system. Creating pooling substation for cluster of RE generators and connecting them with receiving station at appropriate voltage will lead to optimal utilization of transmission system.
- There is also a need for proven high-capacity wind turbines, generally greater than 1–2 MW. In addition, there is a need for turbines to adapt to low-wind regimes and improved design for rotor blades, gear boxes and control systems (Fig. 4.4).

4.2.1.6 Biomass Energy

Biomass energy is the utilization of organic matter and can be used for various applications. It can serve as reliable alternative to diesel. India's non-commercial energy (fuelwood, dung and crop residues) sector is large. As a consequence, emissions of "black carbon"[4] have been identified as significant regional drivers of

---

[4] Black carbon (sometimes referred to as "soot") is small particles produced by the incomplete combustion of fossil fuels, biofuels and biomass. Evidence has emerged in recent years that black carbon from fossil fuels and biomass is second only to carbon dioxide in contributing to climate forcing, and its effect on sensitive areas such as glaciers is even more pronounced. Black carbon resides in the atmosphere for only 1–2 weeks, whereas carbon dioxide remains for hundreds of years. Consequently, major reductions in black carbon emissions can have immediate climate benefits, both regionally and globally. Although black carbon plays a major role in driving regional warming, it is not a "greenhouse gas" and is not covered by the UNFCCC and Kyoto protocol.

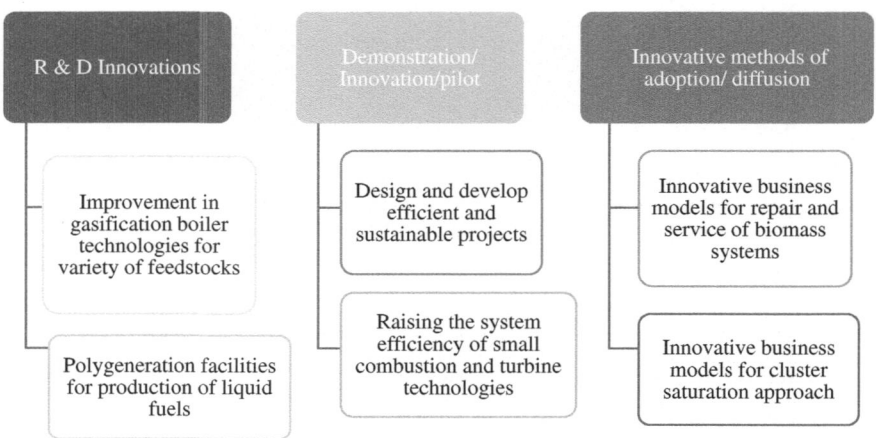

Fig. 4.5 Prioritization options for biomass energy

global warming and cause of serious safety and health problems, including respiratory illness from indoor air pollution; besides, it impacts women and children disproportionately (Fig. 4.5).

MNRE has taken the initiative to provide 1 lakh family-size biogas plants annually. It has launched National Biomass Cook-Stove Initiative to address the issue of inefficiency in biomass combustion including a research programme to identify the right stoves and a pilot project to test the efficiency and marketability of improved community cooking stoves.

Barriers

- One of the most critical bottlenecks for biomass-based plants (based on any technology) is the supply chain bottleneck that could result in non-availability of feedstock. A related problem is price volatility.
- Lack of technical capacity.
- Lack of reliable resource assessment.
- Lack of knowledge about viable and sustainable production and delivery models.

Opportunities for NCEF in Biomass-Based Energy

- Strengthening the activities of MNRE. Biomass-based projects are sensitive to local factors; therefore, multidisciplinary research is needed to design and develop efficient and sustainable projects. There is huge gap in this area.
- Improvement in gasification of various sizes of engines and boiler technology for various feed-stocks in the process.

- Innovative business models for repair and service.
- Innovative business models for cluster saturation approach instead of scattered one for installation of plants and involving local entrepreneurs.
- In bio-energy, opportunities are many and include development of megawatt-scale fluidized bed biomass gasifiers; development of poly-generation facilities for the production of liquid fuels, a variety of chemicals and hydrogen in addition to power production; development of more efficient kilns for charcoal production and pyrolysis of biomass; and raising the system efficiency of small (up to 1 MW) combustion and turbine technologies (Fig. 4.5).

4.2.1.7 Community Solutions

The biggest advantage of renewable energy is for augmenting rural electrification or for providing off-grid energy to remote rural areas. The broad categories of models in rural electrification are listed in Table 4.1.

On the smaller scale, renewable energy can connect with its community roots—when combined with smart grid, the potential is huge. For example, a very high percentage of onshore wind capacity in Germany and Denmark is owned by local communities. Similarly, local ownership was the driving force that created the industry and which has been reflected in the huge take-up of rooftop solar in Germany.

Democratizing renewable energy through local ownership will mean that consumers become producers. This offers the prospect that local business (large and small), hospitals and schools, and the domestic sector enter into arrangements whereby their power and heat is sourced locally (e.g. from waste-to-energy schemes, biomass boilers, PV panels and if suitable wind and hydro).

The possibility is that utility-scale renewable projects may in time be complemented by localized renewable energy (electricity and heat) schemes coupled with smart meters, energy efficiency retrofit, electric vehicle and demand management initiatives. This community energy virtual circle may come to fruition only if utilities and large IT players in many jurisdictions become active participators in its development—even though it challenges centralized models.[5]

In Austin, Texas, the Pecan Street project is using stimulus funding and the municipal-owned utility to pioneer smart grid techniques in USA. The Japan smart community alliance's level of ambition is indicated by a number of domestic demonstration projects, together with international initiatives. Japan's new energy and industrial technology organization (NEDO) is building a smart grid on Hawaii's Maui Island and a project in New Mexico.

---

[5] For example, in the Isle of Wight (UK), IBM Toshiba, SSE, Cable and Wireless and Silver Spring Networks have engaged through a community interest company, Eco-Island, to deliver a smart energy network.

Table 4.1 Categories for rural electrification

Category	Key objective	Nature of demand addressed	Issues/concerns
Grid supply with distribution strengthening	To strengthen distribution and supply by extending grid connection	Industrial, commercial and rural livelihood	Local community participation models in metering, billing and collection activities
Distributed generation with grid backup	Augment grid power availability to the rural areas[a]	Industrial, commercial and rural livelihood	Selection of suitable technologies based on appropriate local resource assessment and local community participation business models
Independent micro-grids with local generation	To provide village or a cluster of villages (or hamlets) with electricity to create an independent self-sufficient generation mini distribution network[b]	Commercial and rural livelihood	Selection of suitable technologies based on appropriate local resource assessment and local community participation business models
Individual home systems	Providing household electrification solution to remote island villages through solar home systems (or any alternate energy source)	Rural livelihood, lighting and heat energy	Selection of suitable technologies based on appropriate local resource assessment and local community participation business models

[a]In particular, in areas with low grid penetration and availability of grid power, stand-alone home systems or a stand-alone distributed generation facility (mini-hydro biomass, other technologies) could provide workable and economically viable solutions

[b]Such stand-alone generation and distribution systems would be particularly viable in remote rural areas where providing grid access as well as management of grid-based systems is technically infeasible or is expensive

The challenge will be to achieve genuine local community participation in both smart grid and renewable. In this, the challenges will be identifying workable models, funding channels and changes required in regulatory practices.

New business models will be required to provide these services both locally and at scale, and from the political economy standpoint, there is good potential for associated public goods benefits and local job creation.

4.2.1.8 Energy Efficiency

Reducing base load energy demand via improvements in energy efficiency is often cited among the least cost options for servicing future energy needs and for tackling emissions. In India, many large energy-intensive industries (e.g. cement, steel) are reported to be already using world's best technology. However, significant energy efficiency gains have been identified in relation to small- and medium-sized industries (SMEs), buildings and appliances and through reducing energy losses in transmission and distribution.

Studies on the demand side of energy consumption have shown that payback period for energy efficiency measures is in the range of 2–8 year. The major barriers are perceived risk, uncertainty about technology, costs of disruption and initial financing. In this context, the 12th Five-Year Plan recognizes the need to set up a special fund with seed capital that will be managed at an arm's length from the government, with the participation of the industry. *NCEF may provide block grants to such a fund in support of activities which will fall in the scope of NCEF's core mandate.*

Energy efficiency in industry and other programmes, such as more efficient lighting, appliances and other "low-hanging" fruits like small hydro, has already received substantial attention in terms of both funds and enabling policy support. However, NCEF may need to play a role in innovation and commercialization of new and emerging technologies in this area too. For example, for the case of small hydro, it would be worth including the initiatives that bring in efficiencies and resource conservation in the value chain of small hydro power equipment's manufacturing as well as those that bring about improvement in efficiencies and reliability of operation and maintenance of small hydro power deployment through incorporation of innovative approaches including those on efficient performance monitoring of remote energy systems. Similarly, innovative ideas and solutions for containing transmission and distribution losses can be supported. Unlike in the case of clean coal and RE which needs a balanced approach for resource allocation between R&D and diffusion, for energy efficiency, the focus of NCEF should be on diffusion (Fig. 4.6).

4.2.1.9 Small Hydro and Biofuels

Technological needs in the SHP sector include technology for direct drive low-speed generators for low-head sources, technology for submersible turbo-generators and technology for variable-speed operation.

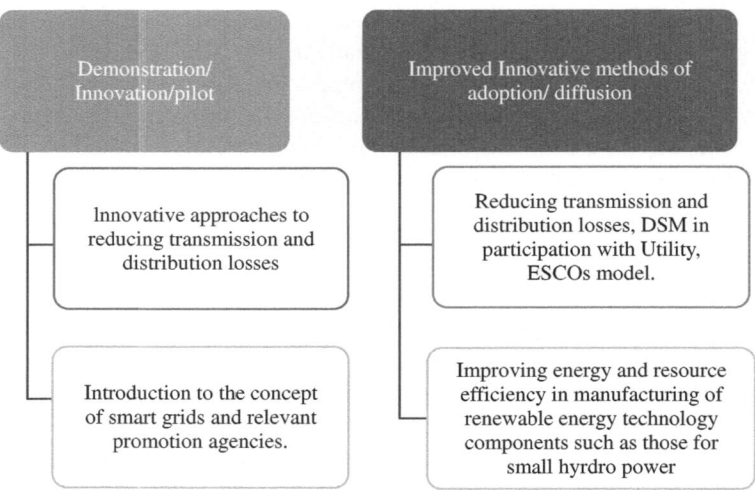

Fig. 4.6 Prioritization options for energy efficiency

Biofuel needs include engine modifications for using more than 20 % bio-diesel as a diesel blend. There is a need for waste-to-energy technological development across the board, including successful demonstration of bio-methanation, combustion/incineration, pyrolysis/gasification, landfill gas recovery, densification and pelletization. In general, a lack of technical expertise exists in installation, operations, maintenance, troubleshooting and other aspects of clean energy implementation.

4.2.1.10 Emerging but Not Proven Technologies

Carbon capture and storage (CCS) is unlikely to be a key technology in India in the near future. The technology itself is still in development stage globally. There has been limited geophysical assessment of potential storage capacity in India. Another important issue in Indian context is that CCS does not accrue any development co-benefits for India.

Further, the central government in its National Action Plan on Climate Change assumes a cautious policy approach to CCS, stating that the cost as well as permanence of storage repositories is still not firm. However, some organizations have commenced dialogue with international organizations regarding CCS, and the government is a member of the Carbon Sequestration Leadership Forum, suggesting that there is interest in investigating the technology further. NCEF could play a role in establishing linkages with international initiatives and other opportunities in this area, ensuring that India is in the loop such that it can both contribute and benefit from further developments in this area.

4.2.2 Financial and Institutional Support for Accelerating Clean Energy Technologies and Innovative Projects

Clean energy funds use a variety of approaches, based on their specific objectives, to support clean energy development. Some of these approaches are as follows:

Investment Model: Under this approach, loans and equity investments are used to support clean energy companies and projects. In many cases, renewable energy businesses find it difficult to obtain financing since traditional financial markets may be hesitant to invest in clean energy. The rationale behind having the state provide initial investment is to bring the renewable energy businesses and the traditional financial markets to a point where investment in renewable energy businesses is sustainable under its own power. An example is the Connecticut Clean Energy Fund (CCEF 2005).

Project Development Model: This approach uses financial incentives, such as production incentives and grants and/or rebates, to directly subsidize clean energy project installation. These funds typically are put in place to help renewable energy be more competitive in the short term by offsetting or lowering the initial capital cost or by offsetting the higher recurring cost of generation. The rationale behind these incentives is that increased market adoption of renewable energy technologies will ultimately drive down the cost of these technologies to a point where, without incentives, they can compete with traditional generation. Examples include California's Renewable Resource Trust Fund and New Jersey's Clean Energy Program (NJCEP 2005).

Industry Development Model: With this approach, states use business development grants, marketing support programmes, research and development grants, resource assessments, technical assistance, consumer education and demonstration projects to support clean energy projects. The rationale behind these programmes is that they will facilitate market transformation by building consumer awareness and demand, supporting the development of a qualified service infrastructure and investing in technological advancement. Examples include Wisconsin's Public Benefit Fund and New Jersey's Clean Energy Program (NJCEP 2005).

Clean energy funds can choose to use more than one model depending upon its objectives.

For instance, New Jersey's Clean Energy Program uses both project development model and industry development model to pursue its objectives.

> Given its mandate, NCEF would need a combination of above mentioned models in designing a framework for financial support.

4.2.2.1 Stages of Innovation and Supporting Financing Mechanisms

To take an innovative idea to its commercial application involves many steps. Basic research leading to a fundamental breakthrough may open up possibilities of applications. R&D is needed to develop conceptual breakthroughs and prove their

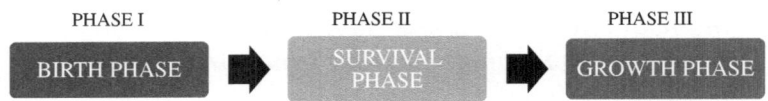

Fig. 4.7 Technological development phases

STAGE 1 R&D	STAGE 2 Demonstration	STAGE 3 Deployment	STAGE 4 Diffusion	STAGE 5 Commercial Maturity

• R&D support

• Grants

• Concessional loans

• Capital subsidy

INCUBATORS
Angel Funds
Product/Process Patenting

• Capital & Production Subsidies

• Fiscal Incentives**

• Concessional/Soft loans

• Venture Capital Funds

• Loan Guarantee

• Public/Private Equity funds

• Insurance

BIRTH PHASE

SURVIVAL PHASE

GROWTH PHASE

**Including creating feed-in tariffs and offering accelerated depreciation, direct subsidies, tax exemptions, reductions on import duty.

Fig. 4.8 Channels of support through various stages of innovation

feasibility. This needs to be followed up by a working, laboratory-scale model. Projects that show economic potential could then be scaled up as pilot projects, while keeping in mind cost reductions that could be achieved through better engineering and mass production. Demonstrations of such projects, economic assessments and further R&D to make the new technology acceptable and attractive to buyers need to be followed, before finally leading to commercialization and diffusion.

An innovative idea translating itself into a successful technological development goes through the phases as shown in Fig. 4.7.

In phase I, an idea gets converted into a workable prototype/process. The next phase is called the "Survival Phase" wherein upscaling of the prototype to the pilot plant/pre-commercial stage is done. In phase III, the pilot production is upscaled to commercial production. Channels of financial and other support through different stages of innovation can be depicted as in Fig. 4.8.

In order to identify opportunities for effective and meaningful intervention by NCEF in this context, it is important to have an understanding of various existing institutional, financial and policy provisions to support and promote research and innovation efforts in India, including those in clean and renewable energy.

Several of the national missions under India's National Action Plan on Climate Change provide a basis for policy measures targeting renewables. In addition, MNRE's Integrated Rural Energy Program (IREP) aims to provide for minimum domestic energy needs for cooking, heating and lighting purposes to rural people in selected village cluster, with a focus on renewable energy. Aside from direct budgetary support, India's main renewable energy financing agency, IREDA, provides finance to renewable energy programmes. Besides, PFC, Rural Electrification Corporation Ltd., NABARD, TDDP and bilateral and multilateral institutions provide funding support at various stages in the supply chain. A summary of various projects/programmes supported by these institutions through various financing instruments is presented in Table 4.2.

In India, R&D and innovation is supported through a number of institutions, programmes and mechanisms (Fig. 4.9). The main barriers and challenges in design, administration and financing of innovation in India are as follows[6]:

(1) Lack of proper system for screening and evaluation of ideas.
(2) Thin support (mechanisms/programmes) to nurture innovative ideas to prototype stage.
(3) Lack of information about (2) as above among potential beneficiaries.
(4) Inflexible funding mechanisms.
(5) Lack of information about innovative products among potential buyers.

A successful clean energy fund will choose a financing model and mechanisms which are designed around the financing gaps and market barriers for the identified R&D and innovation needs in clean energy generation and deployment and supports successful leveraging of private sector funds.

A suggestive framework for financing measures and mechanisms along with policy regimes and the fiscal instruments that go with the various stages of funding by a corpus resource is reflected in a snapshot in Fig. 4.10. This can also be used to assess the environment within which the fund works at several stages of its evolution which could, in turn, be used as criteria for prioritization in the allocation of funds.

4.2.2.2 Framework for Allocation of Funds

The framework for allocation of funds in different projects or sectors will spell out the future course of the NCEF's investments/support. Since the needs of the energy sector in general and clean energy in particular are changing at a fast pace, it will be prudent to keep in mind the time horizons in drawing a road map for sectoral and sub-sectoral fund allocation pattern. For instance, while no one

---

[6] Gupta and Dutta (2005).

Table 4.2 Financial and fiscal support for clean energy

Name of the institution	Objective	Type of financing
IREDA (Indian Renewable Energy Development Agency Limited)	To promote, develop and extend financial assistance for renewable energy and energy efficiency/conservation projects	Finances following kind of projects via 1. Loan with/without rebates in rate of interest, grant or other incentive measures • Off-grid projects under JNNSM • Hydro power projects • Wind energy projects • Biomass cogeneration and industrial cogeneration • Solar photovoltaic/solar thermal[a] grid-connected power projects • Energy conservation/efficiency projects • Projects implemented in ESCO model[b] (with special finances for the electrification of the remote village projects) • Projects eligible for GBI[c] 2. New instrument as loan against securitization of future cash flow of renewable energy project
PFC (Power Finance Corporation)	To extend finance and financial services to promote green (renewable and non-conventional) sources of energy	• Provide term loans to all projects from conventional to renewable energy to government and independent power producers (IPPs) for power generation, transmission and distribution • Other infrastructure projects that have backward linkages to the power sector such as coal mine development, fuel transportation, oil and gas pipelines
REC (Rural Electrification Corporation Ltd.)	To finance and promote rural electrification projects all over the country	1. Provides loans to all entities (government and IPPs) for conventional as well as renewable energy projects 2. Provides short-term loans/medium-term loans and debt refinancing[d] to state power utilities 3. It has set up NEF (National Electricity Fund) as an Interest Subsidy Scheme to promote the capital investment in the distribution sector by providing interest subsidy

(continued)

Table 4.2 (continued)

Name of the institution	Objective	Type of financing
NABARD (National Bank for Agriculture and Rural Development)	To facilitate credit flow for promotion and development of agriculture, small-scale industries, cottage and village industries and handicrafts and other rural crafts. It also has the mandate to support all other allied economic activities in rural areas, promote integrated and sustainable rural development and secure prosperity of rural areas	1. Provides loans, grants or incubation funds[e] to the rural entrepreneurs 2. Under Environmental Promotional Assistance Scheme gives grants and subsidies for undertaking activities related to environment protection aimed at sustainable and environment-friendly agriculture and rural development with a focus on demonstration of replicable eco-friendly technologies
TDDP (Technology Development and Demonstration Program)	To develop and demonstrate innovative need-based technologies for making industry competitive and strengthening the interface between industry, R&D establishments and academic institutions	Provide partial financial support (primarily in the form of grants) covering prototype development, cost of pilot plant, cost of process equipment development, test and evaluation of products, user trials, etc.
Commercial Banks (both public and private sector banks) such as – Bank of Baroda – Bank of Maharashtra – Canara Bank – Bank of India – ICICI Bank – IDBI Bank – Punjab National Bank – State Bank of India – Axis Bank – Indian Overseas Bank – ING Vysya – Laxmi Vilas Bank – Export–Import Bank of India	To allocate financial resources efficiently	• Infrastructure financing • SME financing • End-user financing[f] • Financing the energy equipment for solar energy, bio-energy and clean energy programmes • Transport financing, e.g. green cars • Technology financing • Enables its clients to access global carbon credit market[g]

(continued)

Table 4.2 (continued)

Name of the institution	Objective	Type of financing
Microfinance Institutions such as – Aryavart Gramin Bank – Grameen Surya Bijlee Foundation – Green Microfinance (USA) and Micro Energy International (Germany) launch "Energizing India"[h] with help of the Evangelical Social Action forum – HSBC and Micro Energy Credits – Renewable Energy and Energy Efficiency Partnership (REEP) – TERI (The Energy and Resources Institute) and Clinton Climate Initiative[i] – Self Employed Women's Association (SEWA) – SKS Micro Finance	To entail the provision of financial services to micro-entrepreneurs and small businesses, which lack access to banking and related services due to the high transaction costs associated with serving these client categories	Provide cheap loans (including micro-loans and micro-financing) to • Small/micro-businesses and families to produce micro-energy products and other affordable and portable renewable energy products for poor people • Poor to buy such products • Also funds street lighting systems for villages
Early-Stage Financiers such as – ICICI Technology Finance group – Techno-Entrepreneurship Promotion Program (TePP) – National Innovation Foundation – Centre for Innovation Incubation and Entrepreneurship (RE Search 09 Fund) – The Indian Angel Network	Promote innovation in science and technology and entrepreneurship by providing financial support, constant access to high-quality mentoring, vast networks and inputs on strategy as well as execution	• Providing concessional loans, grants • Cost sharing of the projects • Providing soft loans for unaided, green, grass roots' technological innovation or a traditional knowledge practice
International sources such as – ADB (Asian Development Bank) – World Bank and its private sector arm IFC (International Finance Corporation) – Global Environment Facility (GEF)	Provide financial support and guidance to the domestic bodies	• Private sector financing via equity, loans and guarantees • Technical assistance via special fund • Infrastructure financing • Microfinancing via loan and grants • Setting up of the venture capital fund to support clean technology activities in the private market place • SME financing

(continued)

Table 4.2 (continued)

Name of the institution	Objective	Type of financing
– KfW (Kreditanstalt fur Wiederaufbau), DANIDA (Danish International Development Agency) – USAID (United States Agency for International Development) – Nordic Investment Bank		• Financial support to IREDA, PFC, REC • Providing soft loans[j] and other kind of loans for renewable energy projects

Notes

[a] *Solar thermal electric energy* generation concentrates the light from the sun to create heat, and that heat is used to run a heat engine, which turns a generator to make electricity. *Solar photovoltaic, or PV energy conversion*, on the other hand, directly converts the sun's light into electricity

[b] An energy service company (ECSO) funds the capital expense of energy efficiency improvements upfront, and then, it takes its "interest" or profit in the energy savings achieved over a specific period of time. The ESCO model can be complex, but it is growing in popularity, particularly in the emerging economies. For details, refer http://www.reeep.org/130/esco-model.htm

[c] GBI scheme was introduced by MNRE in 2011 for wind and solar energy projects. Under the scheme for wind and solar energy projects. Under the scheme for wind power, a GBI @ Rs. 0.50 per unit of electricity fed into the grid is provided for a period not less than 4 years and a maximum period of 10 years with a cap of Rs. 62 lakhs per MW. Under the scheme for solar energy, GBI is provided to support small-grid solar power projects connected to the distribution grid (below 33 kV) to the state utilities

[d] Refinancing may refer to the replacement of an existing debt obligation with a debt obligation under different terms. A loan (debt) might be refinanced for various reasons: to take advantage of a better interest rate or a reduced term, to consolidate other debt(s) into one loan, to reduce the monthly repayment amount, etc.

[e] An incubated fund is a fund that is offered privately when it is first created. Investors of this type of fund are usually employees associated with the fund and their family members. Incubation allows fund managers to keep a fund's size small while testing different investment styles before the fund is available to the public and subject to rules and regulations

[f] "End-user finance" is money borrowed by these consumers to pay for energy products or services

[g] Carbon credits and carbon markets are a component of national and international attempts to mitigate the growth in concentrations of GHGs. One carbon credit is equal to one metric tonne of carbon dioxide or in some markets, carbon dioxide equivalent gases. Carbon trading is an application of an emission trading approach. GHGs are capped, and then, markets are used to allocate the emissions among the group of regulated sources. For details of the potential and limits of the global carbon credit markets, refer http://www.gppi.net/fileadmin/gppi/GPPiPP7-Carbon_Markets.pdf. Accessed September 2012

[h] For further information, refer to http://www.microcapital.org/microcapital-story-green-microfinance-of-usa-and-microenergy-international-of-germany-launch-program-to-pro-mote-clean-energy-through-microcredit/. Accessed September 2012

[i] For details, refer http://www.clintonfoundation.org/main/our-work/by-initiative/clinton-climate-initiative/about.html. Accessed September 2012

[j] A loan with no interest or a below-market rate of interest, or loans made by multinational development banks

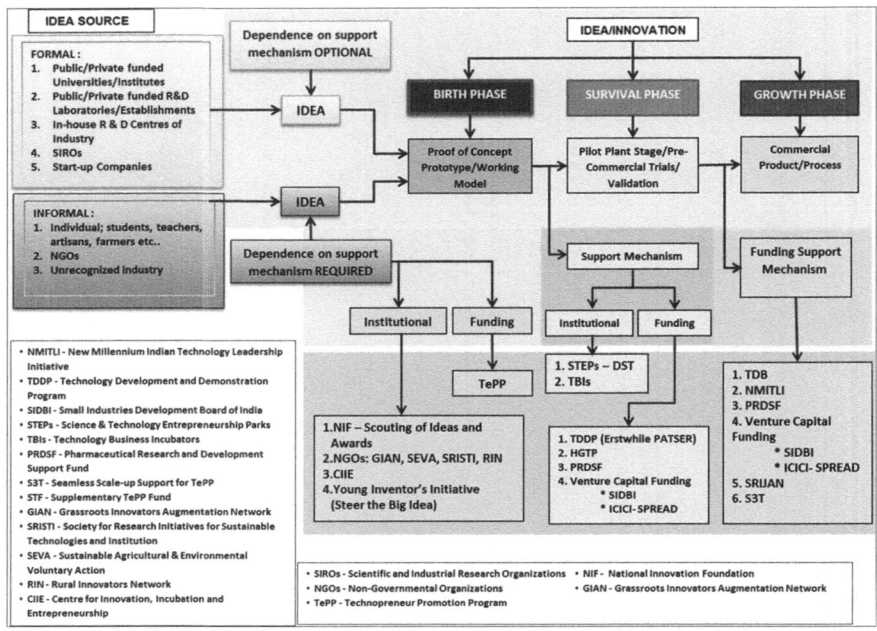

Fig. 4.9 Institutional channels of support for innovation in India.[7] *Source* Gupta and Dutta (2005)

Stage of Development	R&D Innovation	Demonstration	Targeted Deployment	Untargeted Diffusion	Market Independence
Objective	Encourage innovation and entrepreneurship	Prove concept at scale	Support diversity of scalable technologies	Support resource efficiency and competitiveness	Stable and securing ongoing growth
Policy/Tax Mechanisms	• National targets • National research agendas • Fiscal incentives	• Regulatory and legal framework for specific developments/projects • Investment tax incentives	• Feed-in Tariffs • Portfolio standards • Grid development •Targets for specific industries/resources	• Technology neutral renewable energy targets • Carbon taxes • Carbon trading • BAT requirements	With or without: • Carbon taxes • Carbon trading

Examples of Public Finance Mechanisms:
- R&D Grants
- Project Grants
- Guarantees and insurance products
- Soft Loans
- Incubators
- Mezzanine Finance
- Credit Lines
- Public/Private VC Funds
- Public/Private PE Funds
- Technical assistance

Mature Markets
Independent Market

Fig. 4.10 Framework for financing mechanism by stages of activity. *Source* Irbaris and Climate Bonds Initiative (2011)

---

[7] This is not an exhaustive list.

disputes the urgent need for accelerating diffusion of renewable energy, it is also a fact that R&D in renewable energy does not get the kind of attention it deserves for sustaining future growth of renewable energy market at affordable prices.

> In allocation of funds, NCEF may give equal weightage to R&D and demonstration projects (including technology policy and technology road mapping and resource assessment); and to projects for scaling up, deployment and diffusion. Except in the case of energy efficiency projects where deployment and diffusion would need more attention.

To begin with, individuals, academic research institutions, consulting firms and private and public sector enterprise should all compete for this fund. The resources devoted to research in different areas depend on the economic importance of that particular area, the availability of technology and the likelihood of success. The latter changes with time as new developments in science and technology take place and uncertainties reduce.

Financing can be done at various stages, namely pre-development stage that involves investing into R&D and relevant technology; development stage that includes financing the production processes and the training programmes to provide adequate skills; and post-development stage where financing is required to create awareness and marketing of the products/project. Alongside this, various other types of special financing are also required, for instance, infrastructure financing that provides strong forward and backward linkages for the overall growth of the sector.

The fund could include a portfolio of programme options to support both emerging and commercially competitive technologies. Determining both the stage of technology development and the kind of incentives needed to support each technology is an important step in designing a financing model. For emerging technologies, clean energy fund can be used to address a variety of technical, regulatory and market challenges. Technologically proven but relatively expensive solutions will require a completely different approach. For mature technologies that are already cost-competitive, the fund can be used to address other market barriers.

The selection of financial mechanisms and financing tools needs to be programme specific based on a programme's goals. Some financing tools could maximize near-term energy savings and carbon reductions, while others could provide greater funding leverage and long-term impact. The right incentive or tool will depend on that programme's specific goals. Programmes are most successful when leveraging other funding sources.

NCEF in conjunction with other institutions providing support to technology development can play a key role in facilitating a continual evolution of technologies and projects to full commercialization rather than stopgap funding which results in projects falling over at the challenge of moving to the next phase (Table 4.3).

Table 4.3 An illustrative list of financing mechanism for NCEF by type of activity

Activity	Financing mechanism
Technology policy, technology road mapping, other research	Grants (full or part funding depending upon the programme structure)
Resources assessment	Grants; soft loans
R&D, innovation	Grants; soft loans
Technology incubation	Equity; venture capital; soft loan; and grants
Technology demonstration	Grant; soft loan; venture capital; bundling; capital guarantee; risk fund; technology acquisition fund
Innovative methods of adoption/diffusion	Grants; gap finance; soft loan; risk guarantee; equity; support for pooling/blending of technologies

4.2.2.3 Prioritization Across Energy Sectors

The basic guiding principle for sector prioritization has to give due considerations to the efforts that augment the existing national initiatives on

- Inclusive development and energy security to all
- Meeting the commercial energy needs of the unserved population and in providing community-based local solutions
- Research and development of key sectors and technologies
- Building a robust clean energy industry that becomes an important driver of economic growth

Energy security extends to cover issues that diversify the reliable resources of energy on which we develop our energy generation technologies and includes other initiatives that bring about efficiency in the processes by which we extract our energy resources. Coal-based generation is expected to continue to be the predominant source of electricity in the 12th plan period and beyond. Out of the total capacity addition of 75,785 MW envisaged during the 12th plan, coal-based capacity addition is expected to be about 62,695 MW, i.e. about 82.73 %. Energy security issues have to be dealt in close harmony with inclusive development, implying that appropriate consideration be given on making the entire "value chain" of "energy resource" extraction, generation and delivery cleaner and with minimal environmental impacts to the communities and societies living in the vicinity of energy projects. The specific relevance of such efforts will have substantial significance in the case of coal-based power generation.

Integrated Energy Policy points out that it is expected that with a concerted push and a 40-fold increase in their contribution to primary energy, renewables may account for only 5–6 % of India's energy mix by 2031–2032. While this figure appears small, the distributed nature of renewables can provide many socio-economic benefits such as meeting the commercial needs of those in remote rural areas and concurrently augment in expansion of India's domestic energy resource base.

Fig. 4.11 Bandwidth of opportunities for different energy sectors

Major opportunities also exist in reducing energy requirements without reducing energy services. Improvement in energy efficiency or conservation is akin to creating a new domestic energy resource base. Such efficiency improvements can be made in energy extraction, conversion, transmission, distribution and end-use of energy.

In terms of sector prioritization, criterion as mentioned in previous sections certainly creates a pointer of larger "bandwidth of opportunities" for cleaner coal, with renewable energy occupying the next slot followed by energy efficiency (see Fig. 4.11). The bandwidths though indicative of the relative importance are flexible and dynamic; for example, a larger bandwidth may be available for renewables in the following years as more achievements are completed in the cleaner coal sector. The energy efficiency band has the smallest width though it has the larger potential as compared to cleaner coal and renewable energy. This has to take into account the fact that various ongoing measures of energy efficiency are already being undertaken through near-commercial technologies and where line ministries and organizations are very proactive. Therefore, the larger inclusion of energy efficiency initiatives is being seen under those initiatives rather than under NCEF.

The importance of renewables and energy efficiency is duly acknowledged, however given the fact that coal has altogether different challenges to be addressed; therefore, cleaner coal technologies are prioritized independent of renewables. Within the cleaner coal technologies, a suggestive flow of different options is depicted in Fig. 4.2. The priority sectors are at the same levels though the different technological options are prioritized in terms of their respective potential contribution in making the coal value chain clean and resource efficient. The prioritization has been done so as to give a suitable weightage for the private sector participation, especially in the areas where the private sector participation needs to be further strengthened including in areas where private sector is reluctant to enter due to the presence of clear and imminent risks of technological and deployment challenges.

Renewable energy deployment has great potential for augmenting the energy supply options for India using domestic natural resources. The diversity of opportunities for renewable energy is immense owing to factors that are directly related to the large geography over which India's territories extend. The spread of renewable energy resources over such large geography implies that selection of appropriate technological options has to give due consideration to the prevalent local conditions in that geography.

The foremost criterion for wider-scale deployment of appropriate renewable technology has to be based on assessment of the relevant resource potential. A suitable resource potential[8] base that has been firmly validated through scientifically proven reliable methods may be used appropriately for supplementing the links with the agreed areas of "technology development and deployment". The other complimentary criteria for technology prioritization could include the state of technology development, cost, technological adoptability, ease and potential of rapid scale-up, ease of deployment, maintenance skills, infrastructure and other factors.[9]

The contribution of renewable energy will have a critical role not only in providing for the electricity requirement in grid-connected/off-grid mode, but it also has the potential to provide for thermal and cooking needs of variety of end-users including domestic, commercial and industrial. Among such end-users of off-grid energy, a larger group of beneficiaries will be rural and remote population who will have to rely on such renewable energy options for their lifeline needs of cooking and thermal and electrical energy. The long-term benefits of investing in development and deployment of such off-grid technologies could be multiple including associated savings in using conventional sources of energy such as coal-based electricity and fossil fuels (kerosene oil/diesel). A relative measure of such savings could be calculated as defined in Annexure 2.

The intermittent nature of renewable energy technologies for electricity generation presents a difficult challenge for obtaining a higher capacity utilization factor, yet it presents an opportunity to augment energy supply by integrating diverse renewable energy systems such as wind, solar, biomass and small hydro. The challenge in such cases will be integrating and optimizing the generation output from contributing renewable energy technologies in a cost-effective way including through the establishment of a localized micro-grid.

The cost of renewable and other clean technologies is still high, and these have to go through a cost reduction curve before they are ready to compete (without subsidies) with conventional sources of energy. In the interim, the large-scale deployment of renewable energy technologies will be dominated by the technologies that have reached the optimal level of cost reduction, and hence, they are closer to the commercialization than their other counterparts.

The more commercial and mature renewable technologies are areas where the private sector is very active and is the driving force behind the large-scale deployment of such technologies. The result of such active private sector contribution has triggered the growth of large-scale deployment of grid-connected renewable energy technologies such as wind and solar.

The NCEF thus can prioritize off-grid over the grid-connected renewable power generation owing to its larger potential and the issues such as community benefits and being environmentally benign. The focus could be on smaller-scale projects that could be bundled together to achieve larger deployment opportunities. With

---

[8] Under the MNRE-led programs, a detailed resource assessment of solar, wind, biomass, small hydro and other renewable energy technologies has been carried out and validated. Though this has been deemed to be comprehensive, yet technological developments are opening up for new avenues that may enhance the resource potential.

[9] Including the political, economic, socio-cultural, technological and legal factors.

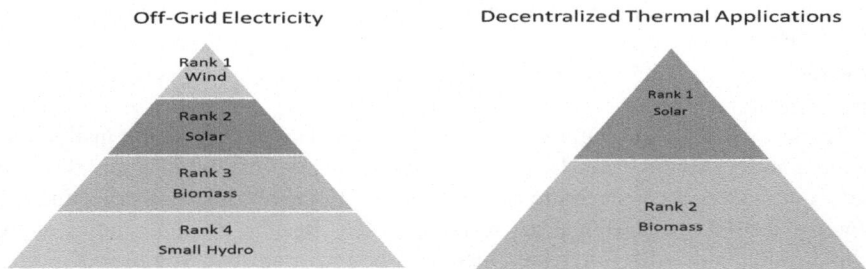

Fig. 4.12 Technology prioritization for off-grid electricity and decentralized thermal applications

that background, the technology prioritizations are indicated in Fig. 4.12. For the off-grid electricity applications, wind energy deployment is prioritized over solar (PV and solar thermal electric), while biomass and small hydro are ranked third and fourth. Similarly, in decentralized thermal application, solar thermal for heating in domestic, commercial and industrial application is prioritized because of large potential. This is followed by bagasse-based cogeneration and biomass gasifiers for thermal applications alongside family-size biogas plants and biomass cookstoves. The competitiveness of renewable and clean energy technology has to be brought about through a balanced and concerted approach that focuses on addressing the gaps on both technological and financial aspects of deployment.

Such approach will have components spread over technology refinements/innovations, market creation and developing supportive manpower and infrastructure.

4.2.3 Skill Development

Skill development will be an important catalyst for sustained growth of clean and renewable energy sector in India. This will be particularly crucial in the case of SMEs, off-grid and community solutions. Since it is difficult for small companies, local governments and community associations/federations to invest in skilling programme, an institutional skilling programme needs to be developed for this segment. Germany's model in institutional skilling can provide a framework in designing a system in India. One of the important issues in this context is how banks and other private sector institutions can be encouraged to be partners in this effort?

With the setting up of the National Skill Development Corporation (NSDC), the government has started to put in place the pieces to train people in the age group of 18–35 years.[10] This is designed to be a demand-driven model.

---

[10] NSDC, which the government set up in 2009 to fund private entities—through loans, equity and grants—to impart hard and soft skills to young Indians for entry-level jobs. Its target is to make 150 million people job-ready by 2022.

Supported by this programme, Noida-based GRAS runs 42 centres in North India. It offers entry-level courses in several sectors, including IT, retail, construction and sales and also imparts soft skills. Companies, too, benefit from this engagement: they get more numbers and better quality.

To start with, NCEF could work through NSDC and other existing institutions in the country. Simultaneously, it should engage expert institutions/individuals to carry out studies for a scientific assessment of the gaps in relevant skills and the efficient institutional mechanisms to address this. Results of such studies along with consultations with the stakeholders will be important building blocks in preparing a strategy for an effective skilling programme.

4.2.4 Knowledge Creation and Sharing

Through collating and providing information on potential, trends, risks, opportunities and best practice, NCEF could be a repository of information as well as a platform to publicize success stories and goals that have been reached. It is important that relevant stakeholders are aware that the clean energy fund is working and achieving the desired results. Knowledge creation will be supported by creating research parks and centres of excellence.

Sharing of lessons and expertise from successful projects/programmes and transfer of knowledge can also help motivate performance and build capacity, thus increasing the effectiveness of the fund manifold.

It is important to develop a stakeholder communication process. Often clean energy funds are established after a robust stakeholder process that includes input from utilities, energy users, equipment manufacturers, project developers, state energy departments and clean energy advocates. However, this has not happened so far in the case of NCEF. A stakeholder process is crucial to ensuring that market realities are given due consideration in both the programme design and implementation process.

In this context, an 'Energy policy, and technology and innovation forum' may be set up which can serve as a platform for recognising and rewarding innovation, and sharing knowledge and best practices. The bigger ambition would be that the important results/best practices feed into political process and international discussions.

4.3 As an Anchor for Establishing Linkages with International Organizations

Combining a range of clean energy programmes and funding within one organization at the national level not only allows for a cohesive strategy for addressing a range of clean energy market issues but also provides a credible platform for developing linkages and cooperation with international clean energy funds, programmes and technical, scientific and other institutions.

As part of the technology road mapping process for a developing country such as India, it would be important to assess whether the foreign collaborations are needed and how foreign linkages and tie-ups can best further the technology strategy and the road map. For example, linkages with appropriate international research organizations and engineering firms might add significant value and speed up basic and applied research for specific technologies. Financial and other logistical support of various bilateral and multilateral organizations can be leveraged in this context. Such arrangements and cooperation may also improve the feasibility of commercial tie-ups and joint venture projects as we move closer to the technology deployment and commercialization phase.

NCEF could also play a role of creating an entry point for potential foreign investors in innovation. It would be important to assess its potential especially in the context of the phenomenon of reverse innovation[11] which is on the rise both as a concept and on the ground. Reverse innovation is any innovation that is adopted first in the developing world. The fundamental driver of reverse innovation is the income gap that exists between emerging markets and the developed countries. The main arguments are as follows:

- Buyers in developing countries demand solutions on an entirely different price–performance curve. They demand new, high-tech solutions that deliver ultra-low costs and "good enough" quality. Thus, adaptation will no longer be an attractive option.
- Thus, developing and poor countries are expected to increasingly become R&D laboratories for breakthrough innovations in diverse fields such as housing, transportation, energy, health care, entertainment, telecommunications, financial services, clean water and many more. If Western multinationals do not innovate for customers in developing countries, they not only stand to lose growth in these countries, but also their competitive position in home markets. This has been seen happen in the 1970s and 1980s when Japanese companies disrupted the Detroit automakers. Emerging giants will do the innovation and bring those innovations into developed countries and disrupt multinationals. We are already seeing strong local players such as Tata, Mahindra, Haier, Lenovo, Goldwind, Suzlon, Cemex and Embraer. The biggest competitors for multinationals are local companies from emerging markets.
- Reverse innovation requires a decentralized, local market focus. Local companies have deep understanding of local customer requirements and problems. But multinationals have deep global capabilities. Both have different strengths to excel at reverse innovation. Perhaps, strategic alliances between local players and multinationals might hold the key.

---

[11] Based on Govindarajan (2012).

- Once tested and proven locally, products developed using reverse innovation must be taken globally, which may involve pioneering radically new applications, establishing lower price points and even cannibalizing higher-margin products. Now more than ever, success in developing countries is a prerequisite for continued vitality in developed ones.

4.4 As an Anchor for Synergy and Linkages with Domestic Institutions

R&D and innovation in the entire supply chain of energy as well as in demand-side management requires strategic and constant interactions between academic researchers, R&D laboratories and industries (manufacturers and utilities) and consumers. In India, this linkage is rather weak or absent in many cases. In the absence of an institutional facilitator and connector, R&D efforts are often not synergistic. NCEF could play the role of a facilitator and connector between relevant stakeholders. Linkages with organizations such as MNRE, IREDA and BEE are also critical to establishing a continuity of financing and keeping a check on unintended overlaps. Convergence among departments/programmes/schemes is also important to avoid thin spread and overlaps.

Further, many states have clean energy funds and/or departments and have their own programmes. NCEF should also develop linkages with state clean energy funds with a view to complement and strengthen each other's efforts.

4.5 A Dedicated NCEF Team with Appropriate Expertise and Accountability

4.5.1 A Professional Organization with Clear Mandate and Accountability

Given the enormous mandate of and expectations from NCEF, it is important that its administration is in a dedicated mission mode. The mission will have the governing, steering and executive arms/groups besides an advisory group at least initially in designing a technology and innovation programme. Ensuring that a fund administrator has access to adequate staffing with appropriate expertise is equally important. Also, rigorous evaluation with clear and consistent metrics and performance targets is essential to shape programme design, motivate performance and monitor results. In other words, the fund will need to be designed, perceived and administered as a professional group/organization with clear mandate and accountability.

4.5.2 Administrative Structure

Based on specific goals and situations, several organizational models for administering clean energy funds have been employed (see Chap. 3). There are examples of specialized institutions being commissioned to administer the clean energy funds. For instance, Massachusetts province in USA chose the Massachusetts Technology Collaborative (MTC) to administer its clean energy fund because MTC's charter, which is to foster high-tech industry clusters in Massachusetts, was consistent with one of the fund's main goals—create a clean energy industry. Also, Connecticut province chose to administer its clean energy fund through Connecticut Innovations Incorporated (CII), a quasi-public state agency charged with expanding Connecticut's entrepreneurial and technology economy. CII's experience in building a vibrant technology community in Connecticut fits well with the challenges of developing a clean energy industry and market.

NCEF may continue to be housed in and administered by MoF. However, it should have adequate and dedicated staffing with appropriate expertise. The process of setting up of the above-mentioned governing, steering and executive arms of NCEF will throw more light on the number and required expertise of the NCEF staff.

Vast experience, expertise and reach of existing institutions such as MNRE, DST and NSDC, and others such as *SRISTI, which nurtures and supports young innovators at regional and grass roots levels*, may be utilized in implementing a wide range of NCEF programmes through programme-based grants. For instance, MNRE has already made inroads in rural electrification, decentralized and community solutions, and technology improvement in solar small appliances. NCEF may choose to either strengthen these programmes if they fulfil the laid criteria or sponsor new programmes. Similarly, DST has the experience of supporting and nurturing innovation through incubators and other such programmes. DST's expertise, experience and institutional set-up can be utilized gainfully to institute similar programmes in clean energy. The Advisory Council of NCEF represented by key stakeholders may further help in identifying appropriate programmes that may be implemented through these institutions. NCEF may also opt for outsourcing some identified activities such as technical review of applications, monitoring and evaluation of projects and programmes to specialized institutions.

The evaluation of the NCEF should be guided by the following principles: independence, transparency, accountability, stakeholder participation, effectiveness and alignment with the principles of the mission statement. It is proposed that independent evaluations be conducted every 2 years, with update reports prepared every year, which should be Web-published and made available to the stakeholders.

References

CCEF (2005) Available on http://www.ctcleanenergy.com/Home/tabid/36/Default.aspx
Chelliah RJ, Appasamy PA, Sankar U, Rita Pandey (2007) Ecotaxes on polluting inputs and outputs. Academic Press, New Delhi

Chikkatur AP, Sagar AD (2007) Clean power in India: towards a clean-coal-technology road-map. Discussion paper 2007–2006, Kennedy School of Government, Harvard University, Cambridge, MA 02138, USA

Financial Express (2013) Pune's solar warriors. Available at http://www.financialexpress.com/news/pune-s-solar-warriors/1075318/2. Last accessed 25 Feb 2013

Govindarajan V (2012) Interview on how reverse innovation can change the world. India Knowledge@Wharton, 29 Mar 2012

Gupta A, Dutta PK (2005) Indian innovation system—perspective and challenges. Technology Exports, vol VII(4). Indian Institute of Foreign Trade, New Delhi

Irbaris & Climate Bonds Initiative (2011) Evaluating clean energy public finance mechanisms, for UNEP sustainable energy finance alliance and its members

NJCEP (2005) Available at http://www.njcleanenergy.com/main/about-njcep/about-njcep

MNRE (2011) Strategic plan for new and renewable energy sector for the period 2011–2017, New Delhi

MNRE (2012) Available at http://www.mnre.gov.in/mission-and-vision-2/achievements/ last Accessed 7 Jan 7 2013

US Department of Commerce, and International Trade Administration (2008) Clean energy: an exporters' guide to India

Chapter 5
Monitoring and Evaluation of the Projects and Programs Supported by the Fund

Existing framework and operations under the NCEF have been analysed in detail in Chaps. 2 and 3 and indirectly through Chap. 4. Several limitations regarding the structure of the NCEF fund have been linked to the absence of clear guidance on precisely defined targets the fund must achieve, and clear roadmaps to realize these targets. Further since there is little mention of any feedback mechanisms regarding disbursement and especially utilisation of the funds, monitoring and performance assessment of the fund as well as the projects it finances is missing from its current structure.

5.1 Monitoring and Performance Assessment

This chapter has addressed this limitation of the NCEF framework and has laid out in extensive detail, steps and measures that ensure a project is designed as per its objectives and monitored through well laid out indicator of performance during its implementation phase. Tools of evaluation are laid out here such that they allow for post project impact assessments too. Further, if external situations change the approach suggested here allows for flexibility during implementation phase by revisiting project activities and allows for modifications in project design and activities, while ensuring none of the higher level objectives and outcomes of the fund is compromised. It is obvious that depending upon the clean energy sector per se and the phase of project intervention that is being implemented, the indicators of performance will vary across the board. The chapter explains why there are no recommended set of "one size fits all" type of indicators and performance norms. It goes on to elaborate the basic system of logic to be used for finding a specific set of indicators and performance norms for a specific project and demonstrates that they can be well laid out within the logical hierarchy of a project design i.e the Logical Framework Approach (LFA).[1]

---

[1] The two terms Logical Framework (LF or Logframe) and the LFA are sometimes confused. The Log Frame is a document; the LFA is a project *design methodology*.

Note: For most purposes the three terms; LFA, ZOPP and OOPP are terms for the same project design and performance monitoring methodology. The terms OOPP and ZOPP mean respectively;

NCEF's mandate provides for two types of innovation in project and programme intervention as per the Finance Minister's speech.

(a) **Innovation in project technology**: Projects that encourage application of new and innovative technology development. Initiatives in this category will typically focus on piloting a concept that has been proved in research or industrial laboratories but not in the field.

(b) **Innovation in the project design and approach**: Projects that use innovative methods to adapt clean energy technology deployment and diffusion. Individual projects in this category will encourage methods of operationalizing the clean energy funds by identifying innovative methods of removing market barriers including policy and regulatory barriers as well as financial barriers for boosting clean energy use in a large scale. Addressing the impediments will ensure effective spurring of technology deployment and efficient end-uses thus ensuring taxpayers and consumers get best value of the cess collection making up the fund.

Once resources have been allocated from the NCEF to a project, its overall performance to meet stated objectives can be best understood by using standard project monitoring devices using a LFA, irrespective of whether the project under consideration belongs to category (a) or (b) above. This chapter identifies issues that concern the monitoring and evaluation of projects and their links to indicators of performance that make up the core of an LFA. The LFA as mentioned briefly earlier is an internationally adopted method practiced for project design and evaluation. Methods to select indicators and how to set up the LFA for both categories of projects i.e. projects using innovative clean energy technologies or projects that showcase innovative methods of adapting and up-scaling clean energy technologies to deliver energy services, are explained here. Steps to monitor and evaluate performances have been explained,[2] using examples of on the ground projects under implementation and illustrative case studies on renewable energy based initiatives. The examples illustrate that the selection of correct, comprehensive, transparent and complete set of applicable indicators is a challenging exercise requiring extensive consultation and inputs from a large number of stakeholders, who are both direct and indirect beneficiaries. The NCEF's application documents and formats give little or no space to this aspect of monitoring of project performances.

Footnote (continued)

Objectives Oriented Project Planning and in German ZielOrientierteProjekPlanung. All three terms refer to a structured meeting process which we will refer to as LFA.

[2] See Table 5.2 for examples of monitoring and performance indicators to be used for different project activities. See Table 5.4 for example of monitoring and performance indicator to be used for Solar Home systems based project. See Annexure 3 for practical, template of a Renewable Energy based Power Generation Project (REPP), under implementation.

Role of monitoring and evaluation of any funding initiative is linked to the initiative's stated objectives, the activities and methods of implementation. In the short run, an evaluator will be best advised to look to infer the *outcomes* of undertaking any set of project *activities*. In the medium term they need to look for the *outputs*, and in long run trace the *impacts* of the project's activities.

Whether project activities lead to the desired outputs which in the longer run have larger outcomes and noticeable impacts are best assessed through performance indicators. The indicators ought to be devised early in the project eligibility phase using responses to queries on the project's characteristics.

Following terminologies make up a monitoring structure and are important in understanding the concept of LFA.

Activities: These are practical time bound actions that the project carries out to deliver the desired project outputs.

Implementation: Includes indications of coordination with other sources of support, effective administration and management, and cost effective operations in the use of public funds.

Output: The goods and services that the project must deliver to achieve the project.

Outcome: The short to medium term behavioural or systemic effects viz: Adoption of new practices, improved institutional competency, and new policies.

Impacts: A fundamental and durable change in the condition of the pre-project/program scenario like lasting improvements in the status of clean energy penetration. OR indications of barriers addressed viz. private finance leverage, industry development, stakeholder satisfaction etc.

Many factors constrain the full achievement of project objectives, including lack of implementation capacity, unrealistic and over ambitious objectives, governance set up of the program and projects, and lack of time and funds. In this context, it is useful to keep in mind the above chain of reasoning that adds up to a comprehensive monitoring tool. They help define indicators of performance at different stages in a project's life starting from immediate term activities to long term impacts.

5.2 Indicators of Assessment

For projects within the clean energy sector usually the following program-level indicators can be devised to assess their performances. These indicators have been developed through research and consultation with project stakeholders across many country projects spread across the globe in 2000–2006[3] and continue to be

---

[3] The *GEF* unites 182 *countries* in partnership with international institutions, civil society, governments and think tanks. Today the *GEF* is the largest public funder of projects to improve the global environment with extensive tools and documented experiences of best practices.

used widely by development practitioners till date. The indicators listed below reflect clean energy use viz. renewable energy, energy efficiency and cleaning of fossil fuel programs particularly for sustainable adoption and market development. Typically, a project design and its activities will need to be able to respond satisfactorily to the questions raised below (making up their eligibility characteristics) and the indicators of performance will be linked to their fulfilment.

5.2.1 Core Indicators

1. Energy production or savings and installed capacities.
2. Technology cost trajectories the project generates.
3. Business and supporting services development that are encouraged through the project.
4. Financing availability and mechanisms that developed around the project's needs.
5. Policy development for the relevant energy sector that addresses some or all of the existing market, technology or financial bottlenecks in the deployment and large scale use of the technology showcased by the project.
6. Awareness and understanding of technologies among stakeholders.
7. Energy consumption, fuel-use patterns, and impacts on end users.

These indicators are static and program evaluators need to monitor them periodically in order to assess changes over time. Some of them are verifiable immediately at the start of a project intervention while others take longer to be verified quantitatively. The indicators reflect both broader trends as well as specific results of the projects; this means plausible linkages between project activities and changes in the indicators need to be established through undertaking evaluation activities.

The seven core indicators can be applied at three levels of organisation:

- At the *project level*, indicators measure a project's direct activities and outputs—the project-level results for which agencies/ministries implementing them are directly responsible. These are the types of indicators generally put forth in project evaluation and supervision reports by those running a project.
- At the local or *state level,* the indicators become state level profiles of project activities, viz. wind energy deployment across wind farms in Maharashtra state or small and medium hydro based clean energy activities in Uttarakhand.
- At the *country level*, indicators become "national profiles" showing national technology, market, and policy trends for energy efficiency and/or renewable energy in a specific country. Linkages can be inferred between direct project results and national trends to show areas of relevance and influence. It is expected that NCEF projects will be designed to influence national trends directly.

A project designer and those involved in assessing funding eligibility will need to mention clearly the level at which a project is to target its activities and hence draw the system boundary of the project's impacts assessment.

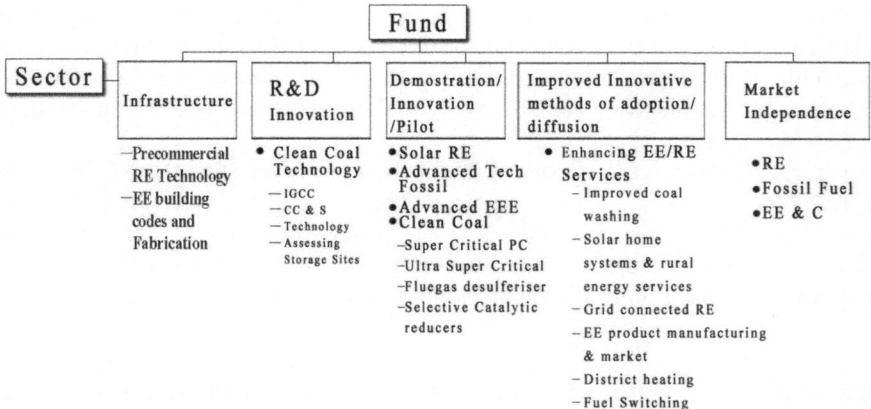

Fig. 5.1   Potential projects across sectors and stages of development

5.3 Project Classification Method to Facilitate Monitoring

For the purpose of putting projects in a monitoring framework one can choose from several project classification methods available among practitioners and agencies that specialise in project formulation and implementation of projects. All such classification methods provide an approach that may be used by agencies to define and categorize technology projects or development projects. The goal is to provide a basis for applying project management practices to delivery of technology projects or large scale diffusion projects for specific purposes. For example, a project classification method may be used by an agency to establish instructions and guidelines for reporting on technology projects to internal partners or for monitoring and scaling up guidelines for different types of technology projects, or for categorization guidelines for whether and how a project satisfies thresholds as required by the mandates of the NCEF.

Implementing bodies/ministries may choose to use the project classification method illustrated below or an alternate classification method from a different source (For example, Bollinger and Wiser (2001) suggest infrastructure model, project development model and investment model). This approach of classification is primarily driven by instruments of public finance viz. tax, tariff, and equity funds etc. which do not blend too well with energy sector performance norms and indicators of progress. We therefore recommend a scheme that combines the Bollinger method to LFA practices used by multi-lateral energy funding bodies for the energy sector. This approach combines the energy sources, their uses and state of maturity of the concept to technology and the status of market development in each of the subsectors of clean energy. Figure 5.1 reflects the types of energy sector classification one can use for NCEF; given the objectives and deliverables of the fund and the documentation available to support the fund's functioning.

It needs no emphasis that a project classification method must be incorporated into management practices based on specific needs of the funding body, as is the case with NCEF. Thus, although project classification methods may vary, the project management practices need to satisfy the guidelines as identified in standard project management practices.

Figure 5.1 presents a summary of the likely project types that the NCEF would typically be looking at, given its objectives. The five sectors correspond to the possible classification one can draw from the Finance Minister's budget speech and have been matched with the state of the currently available technologies, stage of innovation, stages of innovative application techniques etc. for illustration.

5.4 Project Eligibility Criterion to Facilitate Monitoring

Monitoring of a project must be linked to the stated project eligibility criterion that a project's design must have stated upfront. Irrespective of the category a project belongs to, indicators for monitoring the project will be expected to look for eligibility criteria that have been drawn from one or more of the following primary concerns:

- The relevance of project activities to overall country development and growth needs.
- The project's objective.
- The relevance of project objective to the NCEF's objective.
- The identification of the most significant implementation issues in the context of designing indicators.
- The identification of impacts or likely impacts of energy projects (viz. GHG emission reduction) on climate change issues relevant for national and international negotiations.
- Identification of impacts on domestic pollution.
- Identification of factors that influence project's sustainability and replication (viz. market uptake).

An LFA will be seeking answers to the above set of queries to complete a performance analysis of a program or project. Whether project activities lead to the desired outputs which in the longer run have larger outcomes and noticeable impacts are best assessed through indicators that need to be devised using responses to a subset or all of the seven queries mentioned above.

To build an LFA a schematic guideline of steps for undertaking a logical string of vetting a project's progress and success is presented in Fig. 5.2.

The method, known as Review of Outcomes to Impacts (ROtI), is an internationally accepted project evaluation method used often.[4] [There are other approaches,

---

[4] This was developed as part of the Global Environment Facility's Fourth Operational Performance Study (OPS4) by the GEF Evaluation Office and circulated to all 155 member countries then.

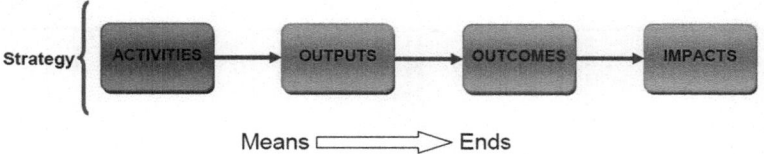

Fig. 5.2 The generic project results chain underlying the theory of change approach. *Source* ROtI Handbook (2009)

developed by the World Bank (10 steps to M and E), USEPA etc. (Kusek and Rist 2004), and use similar logic with varying terminologies.]

The above is a schematic guideline and procedures for undertaking the ROtI, project evaluation method (see footnote 4).[5] The ROtI process uses a Theory of Change approach to evaluate the overall performance of projects (applicable to any environment or energy sector project). It is designed to enable evaluators, through an in-depth analysis of the project's documentation couple with data collection at the project site, to identify and assess the project's component results chains. The results guide the project performance and ultimately contribute to the achievement of project impacts.

Project terminal evaluations are usually conducted at or shortly after project completion, when it is usually only possible to directly assess the achievement of the project outputs and, to a lesser extent, the project outcomes. The long time-frames and lack of long-term monitoring programmes (especially the core funding source that combine additional sources of funding) mean that direct measures of project impacts would require an extensive primary field research that is not possible usually for routine evaluation work. The ROtI's Theory of Change approach seeks to overcome the challenges of measuring impacts by identifying the sequence of conditions and factors deemed important for attribution to the project or program itself.

As can be seen from the diagram, the impact evaluation framework is based on the basic Theory of Change model illustrated in Fig. 5.2, but elaborated to include new components which were felt to be vital to understanding the impact of environmental projects: the intermediate states, assumptions and impact drivers. These three elements are central to the Theory of Change approach adopted in the GEF impact evaluation methodology (as well as the subsequent ROtI methodology), and again are extensively used in national projects of countries signatories to the UNFCCC.

As shown in Fig. 5.3, the impact evaluation methodology developed by the study uses three distinct but complementary analyses for measuring impact, designed to provide a comprehensive understanding of impacts largely based on available project data.

---

[5] UNEP and SEFI, Public Finance Mechanisms To Catalyse Sustainable Energy Sector.

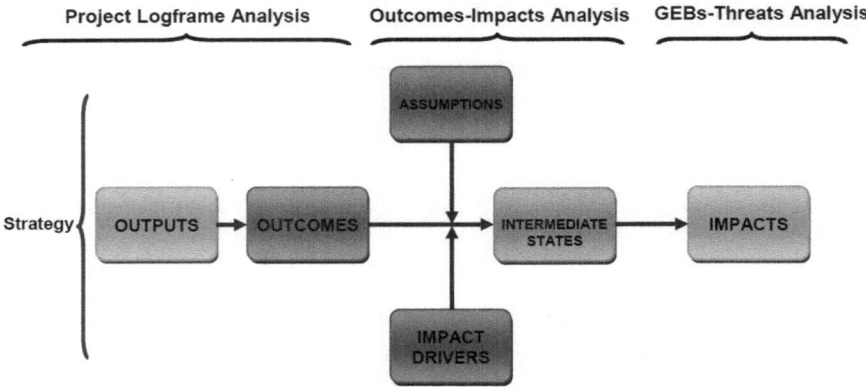

Fig. 5.3 Schematic presentation of project impact assessment framework

The three complementary analyses are:

1. The **Project Logframe Analysis**, which examines the delivery of project outputs and outcomes as defined by the project logical framework.
2. The **Outcomes-Impacts Analysis**, which examines the process by which project outcomes are converted to ultimate impacts through the so-called— intermediate states. This analysis therefore provides a means of indirectly measuring project impacts.
3. The **Benefits/Threats Analysis**, which first identifies the expected project benefits, then assesses project impacts by examining both the change in status of the benefits as well as trends in threats or barriers to these benefits. This is therefore a direct measure of project impacts. While adapting the same logical approach the emphasis for NCEF projects will of course shift to national and local impacts of the project with the global impacts of shifting to a cleaner fuel receiving lower weightage.

The combination of the three different analyses enables the impact evaluation findings to be triangulated, and as a result, the framework provides a relatively robust way of evaluation but is rather time consuming, data intensive and expensive. The Table 5.1 captures the definitions that describe how the three pillars function to give a monitoring and evaluation framework.

Table 5.1 Definitions of different elements of logical hierarchy for the projects

Level	Definition	Examples	Time frame
			Project Life / Short term post project / Long-term
Activities	The practical, time bound actions that the project carries out to deliver the desired project outputs	Construction, communication, training, workshops, research activities, technical advice	
Outputs	The goods and services that the project must deliver in order to achieve the project outcomes. Outputs are within the direct control of the project to deliver	Physical structures, trained individuals, formation of institutions, establishment of service delivery mechanisms, policy instruments and plans, implementation of pilot and demonstration projects	
Outcomes	The short to medium term behavioural or systemic effects that the project makes a contribution towards, and that are designed to help achieve the project's impacts. Achievement of outcomes will be influenced both by project outputs and additional factors that may be outside the direct control of the project	Behavioural changes: adoption of new practices, changed attitudes on issues Systemic changes: improved institutional competency, implementation of new or revised policies, effective decentralising of decision making processes	

(continued)

Table 5.1 (continued)

Level	Definition	Examples	Time frame
Impacts	A fundamental and durable change in the condition of people and their environment brought about by the project. The intended project impacts provide the overall justification for a project. A project will only expect to contribute to the achievement of impact, and often the impact will only be realised many years after project completion	Improved household income, increased environmental resilience. For NCEF: lasting improvements in, and reduced barriers to, the status of adoption of clean energy systems, maintenance and increase in local, national and global environmental benefits	Project Life / Short term post project / Long-term

Source ROtI Handbook (2009)

5.5 Description of Next Steps

5.5.1 Business Justification Review and Process of Selecting Business Solution

Once the energy sector related eligibility conditions of a project are laid out that help in monitoring, standard "good project" characteristics ought to be looked into for ensuring that a rounded monitoring protocol has been established.

Along with following the logical hierarchy of a project a Business Justification Review Gate needs to be clearly spelt out for a project, as it is the initial review gate during project delivery. Business Justification consists of project and/or alternative selection, approval, and initiation. Before a business solution is selected, the agency must examine the solution's investment value in relation to other technology projects and the selection body must assess the project's impact on clean energy resources across the state or the relevant system boundary that it is expected to impact (may be localised, state level or sub regional for example north east India or south western Ghats). Once both these activities have been completed, the proposed business solution may then be formally approved and initiated as a project. Business Justification processes are intended to work in concert with existing implementing body's project management practices.

When a potential opportunity is identified to improve business processes or services through technology, a business case analysis should be initiated. A project's investment value is examined by conducting the business case analysis. The business case analysis compares business case costs to project benefits gained for business process, service, and technology improvements. A key focus is alignment of the project with business goals and objectives. Once completed, the analysis results should help prioritize the project as an agency's, and thus, NCE funds investment.

When a potential opportunity is identified to improve business processes or services through technology, an impact analysis of the project's effect on information resources common throughout the state/states/sub regions must also be initiated. Project impact on use of information technology resources is assessed based on agency responses to an impact analysis questionnaire. The responses can be forwarded to a quality check group of project funding body for review and assessment. The quality check and assurance body must ensure that the proposed business solution does not unnecessarily duplicate existing or on-going efforts and resources are aligned with state/nationwide technology goals and objectives.

If the proposed business solution is selected, the solution should be formally approved and initiated. Approval and initiation of the project signifies that formal project activities can then begin. For example, project roles and staff assignments can then be identified. A critical aspect of initiating a project is refinement of the business goals (if it is not a pure demonstration or pilot project) and objectives identified during the case analysis. A project's primary purpose is to meet the stated goals and objectives.

Table 5.2 Examples of indicators used in conjunction with project activities for monitoring using a project planning matrix

Indicator	Example
1. Energy production or savings and installed capacities	• Number of individual solar home systems installed • Capacity of off grid village power supplies from mini-hydro, biomass, wind, solar
2. Technology cost trajectories	• Installed costs or life cycle system costs of solar home systems • Unit electricity costs of renewable-energy-produced power relative to conventional power costs (e.g., from diesel generators)
3. Business and supporting services development	• Number of solar home system manufacturers, system assemblers, dealers, installers, and service firms (including firms for which solar home systems are not the primary business line) • Existence and appropriateness (to local needs) of equipment quality standards and certification procedures/institutions for equipment and installation
4. Financing availability and mechanisms	• Availability of consumer credit for purchase of solar home systems, including dealer-supplied credit, microfinance, and credit from development banks • Number of financial institutions and volume of lending for off-grid village power
5. Policy development	• Existence of policies and/or plans that explicitly recognize and account for the role of renewable energy technologies in rural electrification • Existence of working regulatory/social models for village power schemes, including tariffs, responsibilities for ownership and maintenance, and equity
6. Awareness and understanding of technologies	• Awareness among rural households of benefits and costs of solar home systems • Abilities of village leaders or project developers to implement and manage village power schemes
7. Energy consumption, fuel-use patterns, and impacts on end users	• Percentage of off-grid households receiving energy services from renewable energy sources relative to conventional sources (by income group or other social parameters) • Consumer satisfaction (by income group or other social parameters)

Before any formal project planning activities can occur (next review gate), the ministry or implementing agency must approve the business outcomes at that specific point during project delivery. Approval indicates project implementers should further invest in delivery of the project.

5.5.1.1 Key Questions for Business Justification

Key questions that must be answered during Business Justification include:

- What business problem does the project solve?
- What other alternatives have been considered?
- What is the impact of not doing this project?
- What is the project's justification, in terms of expected benefits?

Table 5.3 Similarities and difference between the performance indicators of the fund and its projects

Performance criterion for monitoring the fund	Performance criterion for monitoring the projects
1. Thematic alignment of NCE Fund's disbursements to its two core stated objectives as per the FM's speech	The relevance of and thematic alignment of project activities to NCEF's stated objectives
• Support clean energy generation and deployment	Support clean energy generation • Advanced technologies in clean fossil fuel • Advanced technologies in renewable energy including decentralised community
• Support innovation	Support clean energy deployment • Encourage energy conservation and efficiency in use based on innovative delivery mechanisms, especially when technology is already proven and piloted successfully • Encourage RE deployment using innovative technology or service delivery mechanism • Support Innovation in either one/and both as above
2. Thematic alignment of NCE Fund's disbursements to national development objectives	• Project activities are aligned towards either producing more clean energy
• Contributes to encouraging clean energy generation and use without compromising national energy security	• Induce innovative energy service delivery
• Contributes to livelihood and poverty eluviation	• Thereby contribute to improved livelihood and wellbeing • Does not come at a high social or environmental cost (e.g. abatement of local pollution or land-use conflicts), useful acronym: SMART (specific, measurable, attainable, relevant and tractable) • Contributes to global GHG reduction and climate change

- When will the project deliver expected benefits and business outcomes?
- What are the opportunities for reuse of business processes and technical components?
- What was the BASE line situation at the inception of the project?

5.6   Examples of Tools and Deliverables

When a project is going through the business justification phase, several deliverables are completed. Templates and a questionnaire need always be provided as tools at this stage for development of these deliverables and their monitoring at a later stage. Some examples of such instructions are given in Table 5.2.

Table 5.4 Examples of indicators used in conjunction with project activities for monitoring for solar home systems

Indicator	Examples
1. Energy production or savings and installed capacities	• Number of individual solar home systems installed • Capacity of off-grid village power supplies from mini-hydro, biomass, wind, and solar PV (MW)
2. Technology cost trajectories	• Installed costs or life cycle system costs of solar home systems • Unit electricity costs of renewable-energy-produced power relative to conventional power costs (e.g., from diesel generators)
3. Business and supporting services development	• Number of solar home system manufacturers, system assemblers, dealers, installers, and service firms (including firms for which solar home systems are not the primary business line) • Existence and appropriateness (to local needs) of equipment quality standards and certification procedures/institutions for equipment and installation
4. Financing availability and mechanisms	• Availability of consumer credit for purchase of solar home systems, including dealer-supplied credit, microfinance, and credit from development banks • Number of financial institutions and volume of lending for off-grid village power
5. Policy development	• Existence of policies and/or plans that explicitly recognize and account for the role of renewable energy technologies in rural electrification • Existence of working regulatory/social models for village power schemes, including tariffs, responsibilities for ownership and maintenance, and equity
6. Awareness and understanding of technologies	• Awareness among rural households of benefits and costs of solar home systems • Abilities of village leaders or project developers to implement and manage village power schemes
7. Energy consumption, fuel-use patterns, and impacts on end users	• Percentage of off-grid households receiving energy services from renewable energy sources relative to conventional sources (by income group or other social parameters) • Consumer satisfaction (by income group or other social parameters)

5.6.1 Comparison Between Monitoring and Performance Evaluation of a Project and a Fund

While several indicators of the fund's performance and projects' performances would be similar, if not identical, their scales and coverage (system boundaries) will be different. In addition, individual projects would need to be monitored with some additional sector specific or project module based performance indicators that may not be relevant for the overall fund's disbursement. Tables 5.3 and 5.4 list similarities and differences between the performance indicators of the fund and its projects.

References

Bollinger M, LBNL, Bollinger M, Wiser R (2001) Clean energy funds: an overview of state support for renewable energy. Available at http://eetd.lbl.gov/ea/ems/reports/47705.pdf. Last accessed 20 Dec 2012

Kusek J, Rist R (eds) (2004) Ten steps to a result-based monitoring and evaluation system. World Bank, Washington, DC

ROtI Handbook (2009) ROtI handbook: towards enhancing the impacts of environmental projects; Methodological Paper#2, OPS, GEF, 2009

Chapter 6
Evaluation of the Fund

The clean energy sector has experienced significant growth and development globally, especially in the recent years. Dedicated funds have been set up in many countries that support exclusive deployment of clean energy technologies and up scaling of such initiatives through innovative fund disbursement methods and financial support. Countries at varying stages of development and energy mix have been reasonably successful in changing the energy mix in both developing and developed countries. Chapter 3 discussed several funds, ranging from Thailand that launched a petroleum cess-based renewable energy fund to Malaysia, Australia and Canada where dedicated funds function to encourage several forms of clean energy generation and energy services. Several state governments in USA too have dedicated pool of public resources that encourage adoption of clean energy and green energy/renewable energy. Governments around the world have been active in fostering clean energy development to achieve multiple objectives of development and sustainable environment.

Similar reasoning based on the dual objective of economic growth and sustainability has been put forward while setting up a dedicated Clean Energy Fund in India in 2011. Chapter 2 brings out that while mitigation of GHG emission from anthropogenic sources is not a top priority for Government of India,[1] energy security, economic development, employment, and increased energy access are strong justifications in its favour. The same have been noted in Chaps. 2 and 4, constituting the objectives of the NCEF and recalled here to set the stage for the fund's performance evaluation:

(a) encouraging the research and development of innovative clean energy technologies per se (through R&D in innovation and demonstration stages)
(b) supporting innovative methods of adopting clean energy technologies (i.e. targeted deployment and untargeted diffusion)

---

[1] India is a signatory to United Nation's Framework Convention on Climate Change but as a Non-annex 1 country it is not bound by mandatory commitment to emission reduction under the global convention.

Performance of a fund set up to support activities belonging to either or both of the above categories needs to be evaluated at its inception (baseline analysis[2]) and at the end of the term, i.e. the terminal review. Since there is no explicit official reference to an end of term for the NCEF, this chapter will focus on methods of analysing the short-, medium- and long-term performance indicators for the fund, in comparison with the baseline situation. To ensure successful implementation of the fund as per its objectives, not only is it imperative to have mechanisms in place that allows for monitoring of the activities of the corpus fund during its term, but also for midterm course alignment or correction, without impacting the fund's stated objectives. The exercise may be necessary due to unmet or partially met external assumptions or unmet or partially met impact drivers of the fund's resource deployment beyond the control of fund managers or project implementers but would end up impacting the project results nevertheless, unless course corrections are instituted.

6.1 Baseline Identification

Chapter 5 laid out in extensive detail the process to be followed for any project's performance appraisal. In Chap. 6, the emphasis shifts from a project or a programme to the overall performance assessment methods of the clean energy fund itself. To begin with it will be imperative to take steps that demarcate the baseline at the start-up of the fund. The baseline identification will need to clearly identify the current national circumstances reflected through enabling environment, clean energy access points, related finance and investment climate, (amount invested, type of investment, loans, grants, grant programmes, local investments, private equity and/or public market transactions, venture capital, market size expectations, existence or lack thereof), low carbon value chain (availability of local manufacturing and supply chain of clean energy goods, services and financing) and ongoing greenhouse gas management activities at the beginning of fund's activities. Financing conditions in any given country are critical for developers and investors alike. At this stage of development and deployment of the fund, the sovereign cost of debt often serves as a useful benchmark for country risk and is thus an important indicator to be considered by those looking to enter a new market for clean energy producers and services.

---

[2] Methods of demarcating the baseline for the fund in terms of the above parameters is well beyond the scope of this study and we assume at the fund's launch the same were undertaken and documented by the relevant authorities. When an evaluation is undertaken either midstream or at the terminal stage, the baseline benchmarks will be useful to validate the indicators of performance.

6.2 Framework for Performance Evaluation

As discussed for individual projects, performance of the clean energy fund too needs to be assessed through a clearly laid out framework using verifiable indicators. This section focuses on identification and application of such indicators for the performance evaluation of the fund. It presents an analytical structure that makes an assessment of collective performance of the projects and programmes making up the fund's aggregated portfolio of activities.

Verifiable indicators usually consist of performance indicators, financial indicators and sustainability indicators of activities undertaken by the fund. The verifiable indicators are performance parameters that translate objectives into measurable indicators for monitoring and evaluation. These are practised currently in the area of applied M&E tools internationally and considered examples of best practice in project and programme designs. The indicators of success need to be supplemented by clearly stated means of verification that detail the methods for acquiring evidence that objectives are indeed being met.

Following are some core issues assessing working of a fund's and their corresponding indicators as well as several complementary issues that work alongside the core issues. They are based on both quantitative and qualitative indicators for characteristics listed below.

Evaluation of the fund will be expected to focus primarily on the following core issues (see Box 1) assessing whether the fund's disbursement process is in line with its objectives.

Box 1

- Supports innovation and application-oriented research and development in clean energy technology (quantitative), for example coal-bed methane capture and utilization.
- Supports innovation in methods/applications of clean energy deployment (qualitative) and entrepreneurship, for example new delivery mechanisms that address financial barriers or policy barriers or improve efficiency. Engagement of self-help groups, public-private-partnership or CSRs are good examples.
- Sustainability of the programme in the medium and long term, including financial sustainability (both qualitative and quantitative).
- Overall clean energy penetration in total energy balance, contributing to changes in the power sector structure and in the mix of renewable based energy. Diversified sources of energy contributing to the grids and increase in decentralized energy sources: solar PV, biofuels, wind and hydro(quantitative).

- Other socioeconomic impacts (qualitative), including contribution to the long-term solution of country's energy security, positive impacts on skill development, productive uses of energy, employment generation, poverty reduction.
- Does not come at a high social or environmental cost (for instance abatement of local pollution or land-use conflicts). (Both qualitative and quantitative, if environmental costs are internalized.)
- Leverages private sector funds (quantitative).
- Number and types of critical market and financial barriers addressed that are comprehensive and do not create contradictions and market distortions of a new kind (primarily qualitative).
- CO_2/GHG emission reduction: Addresses emerging global climate change concerns over greenhouse gas emissions by encouraging cleaner use of fossil fuels and renewable energy sources (both qualitative and quantitative).

Besides the above checks, an evaluation framework of the NCEF will need to reflect following additional complementary features (see Box 2). These are not specific to an energy fund alone but are relevant for any fund's appraisal, running on public resources.

Box 2

- Whether the fund has followed standard fund allocation principles of a corpus public resource.
- Whether it's working is transparent and efficient (qualitative).
- Whether it is following methodologies and principles used by other professional organizations and technical authorities in the concerned field of expertise (qualitative).
- If the disbursement principles followed standard financial assessment tools of fund allocation (qualitative).
- The fund implementers need to identify the financial costs associated with generating a clean megawatt hour of electricity and the role that financing costs play. Other parameters remaining the same, and the fund needs to choose the most cost-effective option to meet the stated objectives (quantitative).[3]
- Besides being cost-effective in operations, there has to be clear indications of coordination with other sources of support, effective administration and management, in the use of the public fund (qualitative).

---

[3] In short, for an evaluation of the fund the evaluators will need to assure themselves of existence of the core set, the complementary set and the indirect set of indicators of the fund's working processes and performance.

The pointers listed above are basically known as process evaluation indicators that capture the internal dynamics of participating line ministries, institutions, instruments, mechanisms and their management practices. These mark the essential attributes to assess any fund's viability.

All the above-mentioned indicators can be used to assess comprehensive performance characteristics of the fund, using them in equal or different order of weightage (if the need be). While it is not claimed here that the lists are fully exhaustive, they do capture the most relevant bench marks against which meaningful monitoring and evaluation can proceed.

6.2.1 Indirect Indicators

There can be indicators of success that are indirectly generated by NCEF's performance relating to policy regime changes impacting the energy sector viz. demonstrated ability to develop market independence or stable and secure ongoing growth of the sectors supported through NCEF in their nascent stages.

Figure 4.10 lays out the progression of a successful energy fund's development stages: starting with innovations followed by a demonstration stage, it graduates and evolves into targeted deployment of technology within increasingly mature markets where in policy regimes and the fiscal instruments in the working environment facilitate the various stages of funding by a corpus resource all along. Though the snapshot is not directly an M&E tool, it is useful to have it as a backgrounder to assess the environment within which the fund works through several stages of its evolution. Further, it facilitates understanding of a fund's performance when the required supportive policy regime is absent or inadequately operational.[4]

Matching the Figs. 4.10 to 5.1, the evaluator of the NCEF can draw meaningful insights into the performance of the fund for cluster of projects comprising the fund's portfolio for each of the stages. A lack of complementarity between the policy regimes and the fund's disbursement strategies would be a cause for the fund's non-performance or unsatisfactory performance.

6.3 Monitoring of the Fund

A quick survey of international experiences in energy fund management practices throws a rich body of documentation of programme monitoring and performance evaluation experiences and methods. These involve application of the basic principles of the logical framework approach[5] to programme design and evaluation, in

---

[4] It may be recalled here that the final stage of development of the energy sector i.e. market independence and its achievement is not within the scope of the current study.

[5] For details see Chap. 5.

Table 6.1 Relevant performance and success indicators

Relevance	Is the programme valid and pertinent?	• Development issues • Target groups • Direct beneficiaries • GHG emission reduction • Expanding access to energy • Strengthening energy security • Complement mission and niche for several emerging funding and development partners in the country
Performance	What progress is being made by the programme relative to the objectives?	• Effectiveness • Efficiency • Timeliness of inputs and results
Success	What is the programme expected to do to bring about change?	• Impact • Sustainability • Contribution to capacity development

Source European Commission, Brussels report to DGVIII, SMART indicators

varying degrees of disaggregation. The degree of disaggregation for an individual project would of course be very much more detail oriented than that of a fund.

Irrespective of the country or the agency under consideration, all performance indicators involve going over the following three broad generic steps, viz. check relevance, check performance and check success using indicators. When utilized to finance clean energy technologies in particular the objectives, outcomes and outputs of the fund's allocation could be captured using the following table in line with the three steps (Table 6.1).

An example of a more detailed version of the above format will involve drawing up the following disaggregated logical framework analysis of the fund as depicted in the following table (Table 6.2).

Using the kind of indicators designed above, evaluation, measurement and verification (EM&V) can be launched for a fund. These upfront processes and techniques will help to measure and document impact of the NCEF, baseline situation onwards. Once the indicators set is in place, the evaluation process can be set in motion. Evaluation involves retrospectively assessing the performance and implementation of a clean energy fund against the SMART indicators laid out at the fund's start-up stage. Fund's evaluations may include one or more of the following:

- Impact evaluations determine the impacts of the clean energy fund (usually on the energy mix and power generation mix) and co-benefits (such as avoided emissions health benefits, job creation and water savings).
- Process evaluations assess how efficiently a programme was or is being implemented with respect to its stated objectives.
- Market evaluations estimate changes in the market place and thus a programme's influence on encouraging future clean energy activities.

Table 6.2 Example of detailed disaggregated logical framework analysis

Service offering	Performance indicators (Output related)	Impact evaluation (Outcome related)
Whether it's working is transparent and efficient?		
Support to national and local governments in design, implementation and monitoring of clean energy responsive policies, programmes and projects	• Number of explicit steps has been taken to reach out to stakeholders as to the existence and functional norms of the fund • State or local plans consciously plan and coordinate activities in line with the fund	• Increase in the number of institutions, NGOs, individuals/and experts in the field as to the existence and functioning of NCEF • Increase in the number of development and planning initiatives with clean energy at the field level
Whether it is following methodologies and principles used by other professional organizations/technical authorities in the concerned field of expertise (qualitative)		
Process indicator	• Standard business protocols are in place and ready to be implemented	• Streamlined business protocols and disbursement procedures popularized and operationalized
Training	• Number of trained personnel in the field • Strategic choice of incentives and facilities in place for relevant skill formation in financial management targeting RE and decentralized energy	• Increase in the number of players who adopt one or more such practices • Overall skill gap reduced • Positive environment in the market towards new and upcoming clean energy generation
Did the disbursement principles follow standard financial assessment tools of fund allocation		
Benchmarks and harmonized energy sector funding protocols across similar technologies chosen	• Benchmark established with informed stakeholders, viz. line ministries, project developers and manufacturers regarding valuation of benefits and costing tools (e.g. social rate of discount applied where necessary)	• Policies in place in local, state and central government levels on access to efficient funding routes • Private sector policies in place to take on new and additional financial procedures • Uniformity in improved efficiency and to clean energy user costs in daily business practices (cost/Gwh)
Service offering	Performance indicators (output related)	Impact evaluation (outcome related)

(continued)

Table 6.2 (continued)

Service offering	Performance indicators (Output related)	Impact evaluation (Outcome related)
Are the financial costs associated with generating a clean megawatt hour of electricity most cost-effective/innovative other parameters remaining the same (quantitative)		
Is there an awareness of the cost-effective measures/combinations	• Develop road map and identifying cost-effective innovative methods of clean energy options for/with enterprises in India	• Costs of clean energy mg w/hour are indeed the lowest going rate, matches established cost norms
Is there an awareness of innovation in technology?	• Information dissemination and outreach practices by NCEF authorities in place	• Increase in enquiry, exchanges and consultation on innovation in new technology and energy services delivery
Is there awareness towards innovative delivery mechanisms?		• Number of national/state/regional agencies that reflect the same in their plan documents and development budgets
Website and newsletter highlighting evidence and experiences (conceptual, methodological, case studies) on workings and practices of CEFs in other countries		• Number of consultations requests sent out to implementers of similar funds in other countries
		• Level of circulation of newsletters and number of hits on the Web-based links to CEFs
Besides being cost-effective in operations, there are clear indications of coordination with other sources of support, effective administration and management, in the use of the public fund (qualitative)		
Develop road map for coordinated financing	• Ensure a clear road map exits that allows for complementary among activities of different players	• Number of participants/institutions demonstrating ability and willingness to link programme planning and design phases in line with NCEF
Financial sustainability of the fund.	• Ensure the fund's disbursement net reaches diverse clean technologies, scales and players	• While size of core funds increase, per cent share of NCEF to non-NCEF resources come down in the mid- to long term

(continued)

Table 6.2 (continued)

Service offering	Performance indicators (Output related)	Impact evaluation (Outcome related)
Besides being cost-effective in operation, there has to be indications of coordination with other sources of support, especially the private sector and market operators (qualitative)		
Develop road map for promotion of clean technologies outside the public sector		• Number of private sector players/NGOs, CSOs/research labs and autonomous bodies actively experimenting with innovative technologies and demonstration under NCEF
Develop road map for promotion of large-scale adoption of methods of implementing/up scaling and deployment		• Market players, manufacturers, banks an corporates comfortable with NCEF initiatives and playing active role

EM&V establishes the credibility and transparency to clean energy funds by demonstrating that investments in renewable energy generation and innovative delivery mechanisms do indeed provide energy and economic benefits. EM&V provides citizens and decision makers with assurance that funds are being spent prudently. From a purely practical perspective, EM&V can help administrators understand the effectiveness of programme strategies and provide a perspective on what works and what does not. This allows for ongoing improvements in programmes with the goal of maximizing net benefits.

6.3.1 Mechanism to Ensure Monitoring and Periodic Independent Evaluation of the Fund's Performance

It is acknowledged that periodic independent evaluation of the performance of any fund set up with public funds is crucial to attain and maintain its sustainability. It should also be stressed upfront in the fund's mandate that there is a need to separate out the independent periodic evaluation of the fund from day-to-day monitoring activities.

It is important to emphasise here that the evaluation function of the fund should be guided by the following principles: independence, transparency, accountability, stakeholder participation, effectiveness and alignment with the principles of the fund's mission statement. It is proposed that independent evaluations be conducted every 2 years, with updated reports prepared every year, which should be Web-published and made available to the stakeholders.

Chapter 7
Findings and Recommendations

7.1 Context and Objectives

India is one of the fastest growing major economies in the world. This growth is dependent on energy, so maintaining this growth trajectory would require the country to ensure its energy security. It is being increasingly recognized that going forward, the country needs to diversify its primary energy sources and attempt to explore cleaner and renewable sources of and solutions for energy.[1] The arguments are as follows: the need to ensure energy security by reducing dependence on fuel imports, securing development dividends through poverty linkages, GHG emissions and the risk of climate change and health benefits of cleaner and renewable energy and clean energy solutions. India has taken several important measures and has made a steady progress in this direction by putting in place a number of institutions, mechanisms and policies, although a lot remains.

This report is developed in the backdrop of above features in India's energy sector and the recent institution of a dedicated funding facility called the National Clean Energy Fund (NCEF) to address some of these issues. The NCEF

---

[1] The term *clean energy* typically refers to renewable and non-polluting energy sources. Renewable energy is derived from natural resources that can be replenished constantly. Renewable energy takes various forms and includes electricity and heat generated from solar, wind, ocean, hydropower, biomass, geothermal resources and biofuels and hydrogen derived from renewable resources. In addition, certain clean coal technologies and energy efficiency measures also fall under the broad definition of clean energy initiatives. The term *Clean Energy solutions* broadly refers to systems which promote, enhance or advance the energy generation, transport, storage and use so as to reduce the environmental footprint and decrease energy intensity. Such systems include products, services, technologies and regulatory and market-based incentives. These have typically focused on the six key sectors: power; transport; industry; buildings; carbon sequestration; and carbon capture and storage.

instituted in 2010–2011, by levying a clean energy cess on coal produced in India and imported coal at a nominal rate of Rs. 50 per ton, is seen as a major step in India's quest for energy security and reducing carbon intensity of energy. Funding research and innovative projects in clean energy technologies and harnessing renewable energy sources to reduce dependence on fossil fuels constitute the objectives of the NCEF. It is observed that utilization of funds from NCEF has been rather low and disbursements, so far, are aligned more with ongoing programmes/missions of various ministries/departments than with the stated objectives of the fund. This poses potential risk of diluting the focus of NCEF with adverse implications for the much needed research and innovation in clean energy sector in India, especially so, in the absence of any identified targets and prioritization.

This study aims to provide a detailed framework for promoting effective utilization and administration of NCEF. It is hoped that the recommendations of the study will inform the government so that appropriate corrections may be made timely. The outputs of the study will also be useful to hone the strategic thinking on a suitable energy technology policy and an assessment of technology needs besides other barriers in clean energy sector in India.

7.2 Findings and Recommendations

7.2.1 Key Findings from Review of Existing Structure and Operation of NCEF

A review of the NCEF clearly brings out that its present structure and framework for operation need to be sharpened and strengthened to improve its effectiveness and performance. The main points that emerge from review are as follows:

- The NCEF guidelines defining the eligibility of the projects for support are too broad-based. This poses potential risk of diluting the focus of NCEF with adverse implications for research and innovation in clean energy sector in India, especially so, in the absence of any identified targets and prioritization.
- The fund lacks a vision, clearly defined targets, a road map to realize these targets and a feedback mechanism to assess, learn and improve.
- Innovative solutions (whether in technology, business models and financial instruments) require a balance of actions along the innovation chain.

Engaging with diverse stakeholders is critical in identifying such a balance in actions. Although the present framework provides for a mechanism to bring on board the experts and key stakeholders outside of government systems, this opportunity has not been exploited.

- Funding limits and funding mechanism are not at all positioned to leveraging either domestic private investment or international resources and markets. Further, projects' ability to garner funding support from other sources should be rewarded and not penalized by making it ineligible for support from NCEF.

- The type and design of projects received for consideration and the nature of discussion on them in IMG meetings point to an outlook that NCEF can be used freely to fund routine projects and schemes of various ministries as long as they meet a few general requirements. For instance, the discussions have largely focused on what revisions need to be made to a project proposal such that it fits better into the scheme rather than on the merits of the project in terms of its contribution in achieving the objectives of the fund.

- There has been no mention, leave aside a structured discussion that the fund needs to be proactive so as to encourage/invite projects which would promote research and innovation, thus contributing to sustainable development of the clean energy sector.

- The requirement to apply through a central government ministry/department is faulty. Window for direct application should be there.

- Given the objectives of the fund, a dedicated team/mission will be required to administer it. The present structure does not seem adequate and the most appropriate.

7.2.2 Key Lessons from Review of International Clean Energy Funds

The following observations are made on the basis of a review of some identified international clean energy funds. Successful clean energy funds will have the following features:

- A successful clean energy fund will identify its role from other government and non-government programmes and will have focused approach to realizing its objectives. For instance, the Clean Energy Finance Corporation (CEFC) in Australia set up as a mechanism to help mobilize investment in renewable energy, low emissions and energy efficiency projects and technologies has identified a strategy that builds on the existing government grant funding programme for R&D and thus focuses on pre-commercial and commercialization stages in the supply chain. Similarly, the California Clean Energy Fund's (CalCEF) investment strategy is focused on identifying and solving gaps and barriers that are slowing expansion of clean energy markets and adoption of clean technologies.

- A fund having multiple objectives will prioritize its activities and maintain transparent guidelines for allocation of funds among different activities. For instance, CalCEF uses two platforms to run its programmes in Clean Energy, namely CalCEF Capital, for their investment programmes, and CalCEF Innovation, to design and pilot business models, financial products and public policies that grow clean energy markets and accelerate adoption of clean energy technologies. Similarly, CEFC maintains clear guidelines on allocation of funds among renewables, low emissions and energy efficiency projects.
- A successful clean energy fund will constantly engage with diverse stakeholders—leading industry and investment firms, experts, policy makers, academics, scientists, advocates and consumer groups—to get a constant stream of insights into the challenges facing this unique and critical industry. Canadian Green Municipal Fund (GMF) has a multi-stakeholder 15-member Advisory Council and 75-member Peer Review Committee to advise and help and follows an independent third-party technical assessment of proposals. These features of GMF help ensure that projects selected for support are technically sound, besides imparting transparency to its operations, thus improving the overall efficiency of the fund.
- Clearly identified and measurable targets, both quantitative and qualitative, along with a time frame are crucial for a well-functioning clean energy fund. Specific objectives and quantitative targets along with a time frame are set in the Energy Conservation Promotion Fund (ENCON) programme in Thailand. For instance, a Conservation Programme has been developed to provide a guideline for the utilization of the ENCON fund. Three different agencies with relevant expertise have been commissioned to manage different aspects of the programme.
- The fund will have a comprehensive plan to create and share knowledge and build capacity across the country. By strategically allocating funds to the best projects and studies and sharing the lessons and expertise from those initiatives with other municipalities across Canada, effectiveness of GMF increases manifold.

Appropriate administrative structure and access to adequate staffing with appropriate expertise is as important as the design of the programmes implemented by the fund. To implement its energy efficiency programme, ENCON has collaborated with commercial banks, and its ESCO fund is being managed by the professional fund managers. The fund managers proactively work with main target group, SMEs, as a single window facility. Similarly, Massachusetts Province in USA chose the Massachusetts Technology Collaborative (MTC) to administer its clean energy fund because MTC's charter, which serves as a catalyst for growing the state's innovation economy, was consistent with one of the fund's main goals—create a clean energy industry. Also, Connecticut Province chose to administer its Clean Energy Fund through Connecticut Innovations Incorporated (CII), a quasi-public state agency charged with expanding Connecticut's entrepreneurial and technology economy. CII's experience in building a vibrant technology community in Connecticut fits well with the challenges of developing a clean energy industry and market.

7.2.3 Proposed Framework for NCEF

A review of NCEF brings out that the actual disbursements/approvals so far from NCEF are aligned more with ongoing programmes/missions of various ministries/departments than with the stated objectives of the fund. Also, utilization of funds from NCEF has been rather low due to the fact that it has been far from being proactive in both identifying appropriate programmes and building a strategy in operationalizing them. However, this is an expected outcome in the absence of a well-thought-out framework for administering the fund. In this context, the following framework is proposed.

- Niche for the fund and value addition of the fund need to be spelt out so that it is properly understood by the stakeholders
- NCEF as an anchor for establishing linkages and cooperation with international institutions/programmes in areas of core mandate of NCEF
- NCEF as an anchor for synergy between other government efforts in areas of core mandate of NCEF
- Dedicated NCEF team with appropriate expertise and accountability

Each of these has been discussed in detail in Chap 4.

7.2.4 Framework for Allocation of Funds

Given its objectives and the proposed framework in realizing these objectives, the core constituencies of NCEF would be as follows:

1. Encouraging the *development* of innovative clean energy technologies per se (through R&D in innovation and demonstration stages).
2. Supporting innovative *methods* of *adopting* clean energy technologies (i.e. targeted deployment and untargeted diffusion).

Hence, the fund can support both types of initiatives either sequentially or choose to support both with equal/unequal weightage.

NCEF should support both innovation in clean energy technology as well as ways of adoption/deployment of clean energy technologies that may have been piloted but await innovative application supporting market creation and deployment.

The framework for allocation of funds in different projects or sectors will spell out the future course of the NCEF's investments/support. Since the needs of the energy sector in general and clean energy in particular are changing at a fast pace, it will be prudent to keep in mind the time horizons in drawing a road map for sectoral and sub-sectoral fund allocation pattern. For instance, while no one disputes the urgent need for accelerating diffusion of clean energy, it is also a fact that R&D in clean energy does not get the kind of attention it deserves for sustaining future growth of renewable energy market at affordable prices.

In allocation of funds, NCEF may give equal weightage to R&D and demonstration projects (including technology policy and technology road mapping and resource assessment); and to projects for scaling up, deployment and diffusion. Except in the case of EE Projects where greater focus would be on deployment and diffusion.

To begin with, individuals, academic research institutions, consulting firms and private and public sector enterprises should all compete for this fund. Creation of research parks, incubation centres and centres of excellence is an accepted practice to go about it. The resources devoted to research in different areas depend on the economic importance of that particular area, the availability of technology and the likelihood of success. The latter changes with time as new developments in science and technology come up and uncertainties reduce.

Financing can be done at various stages, namely pre-development stage; development stage; and post-development stage where financing is required to create awareness and marketing of the project. Alongside this, various other types of special financing are also required such as infrastructure financing which provides strong forward and backward linkages for the overall growth of the sector.

The fund could include a portfolio of programme options to support both emerging and commercially competitive technologies. Determining both the stage of technology development and the various incentives to support each technology is an important step in designing a financing model. *Since there are huge gaps in early-stage funding, NCEF may consider this to be a focus area in allocation of funds.*

R&D in the energy sector is critical to augment and diversify our energy resources and to promote energy efficiency. *The first critical priority would be a suitable energy technology policy and an assessment of technology and innovation needs.* The technology road mapping can add substantial value to the technology policy. This exercise will be based on a dynamic strategic vision and conducted in collaboration with relevant stakeholders.

The next step would be a mapping of various ongoing efforts both institutional and others. This will help NCEF in determining its role from other existing programmes, thus checking overlaps and maintaining focus of different initiatives/programmes, thereby enhancing the overall effectiveness of various initiatives. *The gaps so identified will guide the fund's clean energy technology and innovation programme.*

Innovation is crucial in non-technology aspects of promoting clean energy, namely innovation in supporting policies (regulatory, fiscal), financial products, business models and community-level solutions.

The NCEF should encourage and fund such studies in a number of institutions on long-term basis and should also commission studies to independent experts and consultants. A number of academic institutions should be developed as centres of excellence in energy research. Besides, coordinated research in all stages of innovation chain should be supported. *One of the criteria for selection of centres of excellence would be application orientation of their research and innovation and linkages with regional-/local-level institutions.*

7.2.5 Prioritization Across Energy Sectors

The basic guiding principle for sector prioritization has to give due considerations to the efforts that augment the existing national initiatives on

- inclusive development and energy security to all;
- meeting the commercial energy needs of the unserved population and in providing community-based local solutions;
- research and development of key sectors and technologies;
- building a robust clean energy industry that becomes an important driver of economic strength.

In terms of sector prioritization, criterion as mentioned above creates a pointer of larger "bandwidth of opportunities" for cleaner coal technologies, with renewable energy occupying the next slot followed by energy efficiency as depicted in Fig. 4.11 (see Chap. 4). The bandwidths though indicative of the relative importance are flexible and dynamic; for example, a larger bandwidth may be available for renewables in the following years as more achievements are completed in the cleaner coal sector. The energy efficiency band has the smallest bandwidth though it has the larger potential as compared to cleaner coal and renewable energy. This has to take into account the fact that various ongoing measures of energy efficiency are already being undertaken through near-commercial technologies and where line ministries and organizations are very proactive. Therefore, the larger inclusion of energy efficiency initiatives is being seen under those initiatives rather than under NCEF.

Opportunities in each of these sectors are discussed in detail in Chap. 4. The main points that emerge from this discussion are summarized in what follows.

7.2.5.1 Opportunities for NCEF in Clean Coal Sector

The importance of renewables and energy efficiency is duly acknowledged, however, given the fact that coal offers altogether different challenges to address;

cleaner coal technologies are prioritized independent of renewables. Within the cleaner coal technologies for a suggestive flow of different options, see Fig. 4.2. Different technological options are prioritized in terms of their respective potential contribution in making the coal value chain clean and resource efficient. R&D in coal mining is minimal in India. Development and adaptation of technologies for mining low-ash coal and efficient coal handling have huge potential in improving resource efficiency and reducing climate change concerns. Coal beneficiation improves its thermal efficiency and reduces operation and transport costs of power plants and other users. Coal bed methane and underground coal gasification are other areas which would need support with technology adaptation.

The adoption and success of supercritical technology will depend largely on the coal quality and its assured supply. The experience in manufacturing, supply and operations of high-pressure and high-temperature main plant equipment is limited in the country. The longer-term impact of higher-pressure and higher-temperature profiles on the boiler and related components' life is not yet known, and closer collaboration between technology suppliers and generators will become important initially. Innovative environment management projects around the coal mines, coal washeries, rail sidings and other coal utilizing plants are also important.

7.2.5.2 Opportunities for NCEF in Renewable Energy

Renewable energy deployment has great potential for augmenting the energy supply options for India using domestic natural resources. The diversity of opportunities for renewable energy is immense owing to factors that are directly related to the large geography over which India's territories extend. The spread of renewable energy resources over such large geography implies that selection of appropriate technology options has to give due consideration to the prevalent local conditions in that geography. A suggestive list of opportunities for NCEF in solar, wind and biomass is depicted in Figs. 4.3–4.5.

The foremost criterion for wider-scale deployment of appropriate renewable technology has to be based on the assessment of the relevant resource potential. A suitable resource potential base that has been firmly validated through scientifically proven reliable methods may be used appropriately for supplementing the links with the agreed areas of "technology development and deployment". The other complimentary criteria for technology prioritization could include the state of technology development, cost, technological adaptability, ease and potential of rapid scale-up, ease of deployment, maintenance skills, infrastructure and other factors.

The contribution of renewable energy will have a critical role not only in providing for the electricity requirement in grid-connected/off-grid mode, but it also has the potential to provide for thermal and cooking needs of variety of end-users including domestic, commercial and industrial. Among such end-users of off-grid energy, a larger group of beneficiaries will be from rural and remote population who would have to rely on such renewable energy options for their lifeline needs of cooking, thermal and electrical energy. The long-term benefits of investing in development and deployment of such off-grid technologies could be multiple including associated savings in using conventional sources of energy such as coal-based electricity and fossil fuels (kerosene oil/diesel).

The intermittent nature of renewable energy technologies for electricity generation presents a difficult challenge for obtaining higher capacity utilization factor, yet it presents for an opportunity to augment energy supply by integrating diverse renewable energy systems such as wind, solar, biomass and small hydro. The challenge in such cases will be integrating and optimizing the generation output from contributing renewable energy technologies in a cost-effective way including through the establishment of a localized micro-grid.

> The NCEF thus can prioritise off-grid over the grid-connected renewable power generation owing to its large potential and favourable factors such as community and environmental benefits. The focus could be on smaller projects that could be bundled together to achieve larger deployment opportunities.

7.2.5.3 Opportunities for NCEF in Energy Efficiency

Reducing base load energy demand via improvements in energy efficiency is often cited among the least cost options for servicing future energy needs and for tackling emissions. In India, many large energy-intensive industries (e.g. cement, steel) are reported to be already using world's best technology. However, significant energy efficiency gains have been identified in relation to small and medium-sized industries (SMEs), buildings and appliances and through reducing energy losses in transmission and distribution.

Studies on the demand side of energy consumption have shown that payback period for energy efficiency measures is in the range of 2–8 years. The major barriers are perceived risk, uncertainty about technology, costs of disruption and initial financing. In this context, the Twelfth Five-Year Plan recognizes the need to set up a special fund with seed capital that will be managed at an arm's length from the government, with the participation of the industry. *NCEF may provide block grants to such a fund in support of activities which will fall in the scope of NCEF's core mandate.* For a suggestive priority list of energy efficiency activities for support from NCEF, see Fig. 4.6.

Energy efficiency in industry and other programmes, such as efficient lighting, appliances and others like small hydro, have already received substantial attention in terms of funds and enabling policy support. However, NCEF may need to play a role in innovation and commercialization of new and emerging technologies in this area too. For example, for the case of small hydro, it would be worth including the initiatives that bring in efficiencies and resource conservation in the value chain of small hydropower equipment manufacturing as well as those that bring about improvement in efficiencies and reliability of operation and maintenance of small hydropower deployment through incorporation of innovative approaches including those on efficient performance monitoring of remote energy systems. Similarly, ideas and solutions for containing transmission and distribution losses can be supported.

Technological needs in the SHP sector include technology for direct-drive low-speed generators for low-head sources, technology for submersible turbo-generators and technology for variable-speed operation. Similarly for biofuels, technology needs to include engine modification for using more than 20 % biodiesel as a diesel blend. There is a need for waste-to-energy technological development across the board, including successful demonstration of bio-methanation, combustion/incineration, pyrolysis/gasification, landfill gas recovery, densification and pelletization.

7.2.5.4 Emerging but not Proven Technologies

Carbon capture and storage is unlikely to be a key technology in India in the near future. The technology itself is still in a nascent stage globally. There has been limited geophysical assessment of potential storage capacity in India. Another important issue in Indian context is that CCS does not accrue any development co-benefits for India.

Further, the central government in its National Action Plan on Climate Change assumes a cautious policy approach to CCS stating that the cost as well as permanence of storage repositories is still not firm. However, some organizations have commenced dialogue with international organizations regarding CCS, and the government is a member of the Carbon Sequestration Leadership Forum, suggesting its interest in investigating the technology further. *NCEF could play a role in establishing linkages with international initiative and other opportunities in this area, ensuring that India is in the loop such that it can both contribute and benefit from further developments in this area.*

7.2.6 Financing Models and Mechanisms

Clean energy funds use a variety of approaches, based on their specific objectives, to support clean energy development. Some of these approaches are as follows: investment model; project development model; and industry development model.

> Given its mandate, NCEF would need a combination of project development and industry development models in designing a framework for financial support.

A suggestive framework for financing mechanisms by various stages of activity through a corpus resource is reflected in a snapshot in Fig. 4.10 (see Chap. 4). This

Table 7.1 An illustrative list of financing mechanisms for NCEF by type of activity

Activity	Financing mechanism
Technology policy, technology road mapping and other researches	Grants (full or part funding depending upon the programme structure)
Resources assessment	Grants; soft loans
R&D and innovation	Grants; soft loans
Technology incubation	Equity; venture capital; soft loan; and grants
Technology demonstration	Grant; soft loan; venture capital; bundling; capital guarantee; risk fund; and technology acquisition fund
Innovative methods of adoption/diffusion	Grants; gap finance; soft loan; risk guarantee; equity; and support for pooling/blending of technologies

can also be used to assess the environment within which the fund works at several stages of its evolution which could, in turn, be used as criteria for prioritization in allocation of funds.

The selection of financial mechanisms and financing tools needs to be programme specific based on a programme's goals. Some financing tools could maximize near-term energy savings and carbon reductions, while others could provide greater funding leverage and long-term impact. The right incentive or tool will depend on that programme's specific goals. Programmes are most successful when leveraging other funding sources.

NCEF in conjunction with other institutions providing support to technology development can play a key role in facilitating a continual evolution of technologies and projects to full commercialization rather than stop-gap funding which results in projects falling over at the challenge of moving to the next phase. An illustrative list of financing mechanisms by type of activity is in Table 7.1.

- NCEF may implement some of its programmes through existing institutions such as MNRE, Bureau of Energy Efficiency (BEE), Ministry of Coal, Department of Science and Technology (DST), Central Mine Planning and Design Institute Limited (CMPDIL), National Innovation Foundation and Sristi.
- NCEF may also consider supporting ongoing programmes, which fall in the scope of its core mandate, of such institutions with a view to strengthening them.
- NCEF may evolve a criteria for selection of these institutions—such as capacity in leveraging private sector and international funds; use of innovative business model; and innovative financing instruments.
- In implementing certain programmes, there may be a need to evolve/set up new institutions in public–private partnership.
- NCEF may make strategic grants to support the launch and growth of important institutions advancing the broad NCEF agenda.
- Matching or proportionate contribution from NCEF to State energy funds for furthering the NCEF mandate has merit and may be considered.

7.2.7 Skill Development

Skill development will be an important catalyst for sustained growth of clean and renewable energy sector in India. This will be particularly crucial in the case of SMEs, off-grid and community solutions. Since it is difficult for small companies, local governments and community associations/federations to invest in skilling programmes, an institutional skilling programme needs to be developed for this segment. An important issue in this context is how banks and other private sector institutions should be encouraged to be partners in this effort.

With the setting up of the National Skill Development Corporation (NSDC), the government has started conjoining pieces to train people in the age group of 18–35 years. This is designed to be a demand-driven model. These initiatives are certainly welcome, but they remain piecemeal responses to an underlying failure to match skill production to skill requirements even in publicly funded schemes such as health and education.

> To start with, NCEF could work with NSDC and other relevant institutions in the country. Simultaneously, it should engage expert institutions/individuals to carry out studies for a scientific assessment of the gaps in relevant skills in clean energy sector, and the efficient institutional mechanisms to address this. Results of such studies along with consultations with the stakeholders will be important building blocks in preparing a strategy for an effective skilling programme.

7.2.8 Knowledge Creation and Sharing

Through collating and providing information on potential, trends, risks, opportunities and best practice, NCEF could be a repository of information as well as a platform to publicize success stories and goals that have been reached. It is important that relevant stakeholders are aware that the clean energy fund is working and achieving the desired results.

Sharing of lessons and expertise from successful projects/programmes and transfer of knowledge can also help motivate performance and build capacity, thus increasing the effectiveness of the fund manifold.

A dynamic and vibrant stakeholder communication process is crucial to also ensuring that market realities are given due consideration in both the programme design and implementation processes.

> In this context, an 'energy policy, technology and innovation forum' may be set up which can serve as platform for recognising and rewarding innovation, and sharing knowledge and best practices. The bigger ambition would be that the important results/best practices feed into political process and international discussions.

7.2.9 Anchor for Establishing Linkages with International Organizations

Combining a range of clean energy programmes and funding within one organization at the national level not only allows for a cohesive strategy for addressing a range of clean energy market issues but also provides a credible platform for developing linkages and cooperation with international clean energy funds, programmes and technical, scientific and other institutions.

As part of the technology road mapping process for a developing country such as India, it would be important to assess whether the foreign collaborations are needed and how foreign linkages and tie-ups can best further the technology strategy and the road map. For example, linkages with appropriate international research organizations and engineering firms might add significant value and speed up basic and applied research for specific technologies. Financial and other logistical support of various bilateral and multilateral organizations can be leveraged in this context. Such arrangements and cooperation may also improve the feasibility of commercial tie-ups and joint venture projects as we move closer to the technology deployment and commercialization phase.

> NCEF could also play a role of creating an entry point for potential foreign investors in innovation.

It would be important to assess its potential especially in the context of the phenomenon of reverse innovation which is on the rise both as a concept and on the ground. Reverse innovation is any innovation that is adopted first in the developing world. The fundamental driver of reverse innovation is the income gap that exists between emerging markets and developed countries.

7.2.10 Anchor for Synergy and Linkages with Domestic Institutions

R&D and innovation in the entire supply chain of energy, as well as in demand-side management, require strategic and constant interactions between academic researchers, R&D laboratories, industry (manufacturers and utilities) and consumers. In India, this linkage is rather weak or absent in many cases. In the absence of an institutional facilitator and connector, R&D efforts are often not synergistic. NCEF could play the role of a facilitator and connector between relevant stakeholders. Linkages with organizations such as MNRE, IREDA and BEE and many others are critical to establishing a continuity of financing and keeping a check on unintended overlaps.

Many States have clean energy funds and/or departments and have their own programmes. NCEF should also develop linkages with State clean energy funds with a view to complement and strengthen each other's efforts. Matching or proportionate contribution from NCEF to State energy funds for furthering the NCEF mandate merits consideration.

Often clean energy funds are established after a robust stakeholder process that includes input from utilities, energy users, equipment manufacturers, project developers, state energy offices and clean energy advocates. This has not been done so far in the case of NCEF; thus, formal channels for meaningful interactions among relevant institutions are even more important.

7.2.11 A Dedicated NCEF Team with Appropriate Expertise and Accountability

7.2.11.1 A Professional Organization with Clear Mandate and Accountability

Given the enormous mandate of and expectations from NCEF, it is important that it is administered in a dedicated mission mode. The mission should have the governing, steering and executive arms/groups, besides an advisory group, at least initially, for designing a technology and innovation programme. Ensuring that a fund administrator has access to adequate staffing with appropriate expertise is equally important. Also, rigorous evaluation with clear and consistent metrics and performance targets is essential to shape programme design, motivate performance and monitor results. In other words, the fund will need to be designed, perceived and administered, as a professional group/organization with clear mandate and accountability.

7.2.11.2 Administrative Structure

Based on specific goals and situations, several organizational models for administering clean energy funds have been employed. There are examples of specialized institutions being commissioned to administer the clean energy funds. For instance, Massachusetts Province in USA chose the Massachusetts Technology Collaborative (MTC) to administer its clean energy fund because MTC's charter, which is to foster high-tech industry clusters in Massachusetts, was consistent with one of the fund's main goals—create a clean energy industry. Also, Connecticut Province chose to administer its Clean Energy Fund through Connecticut Innovations Incorporated (CII), a quasi-public state agency charged with expanding Connecticut's entrepreneurial and technology economy. CII's experience in building a vibrant technology community in Connecticut fits well with the challenges of developing a clean energy industry and market.

NCEF may continue to be housed in and administered by MoF. However, it should have adequate and dedicated staffing with appropriate expertise. The process of setting up of the above mentioned governing, steering and executive arms of NCEF will throw more light on the number and required expertise of the NCEF staff.

Vast experience, expertise and reach of existing public sector institutions such as MNRE, DST, NSDC and others such as *Sristi*, which nurtures and supports young innovators at regional and grass root levels, may be utilized in implementing a wide range of NCEF programmes through programme-based grants. For instance, MNRE has already made inroads in rural electrification, decentralized and community solutions and technology improvement in solar small appliances. NCEF may choose to either strengthen these programmes if they fulfil the laid criteria or sponsor new programmes. Similarly, DST has the experience of supporting and nurturing innovation through incubators and other such programmes. DST's expertise, experience and institutional set-up can be utilized gainfully to institute similar programmes in clean energy. The Advisory Council of NCEF represented by key stakeholders may further help in identifying appropriate programmes that may be implemented through these institutions. NCEF may also opt for outsourcing some identified activities such as technical review of applications, monitoring and evaluation of projects and programmes.

It is proposed that independent evaluations of the NCEF be conducted every 2 years, with updated reports prepared every year, which should be Web-published and made available to the stakeholders.

Information system must improve. Information on structure, framework, application procedure, activities and achievements should be Web-published, constantly updated and made available to the stakeholders.

7.2.12 *Monitoring and Evaluation of Activities Supported by NCEF*

The study recommends that a monitoring protocol be put in place for every project/programme, keeping the identified logical links between the objective of a project, its eligibility criterion, the activities and the outcomes in view. The report has laid out in extensive detail measures that can be used to ensure that projects and programmes are designed per its objectives and monitored through well-laid-out indicator of performance in short- as well as long-term horizon during its implementation phase. Methods and tools of evaluation are also laid out to allow for post-project impact assessment. Further, if external situations change, the approach allows for flexibility during the implementation phase by revisiting project activities and allowing for modification, but ensuring none of the higher-level objectives and outcomes are compromised. It is obvious that depending upon the clean energy sub-sector and phase of project intervention being implemented, the indicators of performance will vary across the board. However, the basic logic behind finding the specific set meant to be used for specific projects is well laid out within the logical hierarchy of the Logical Framework Approach (LFA) to project monitoring.

7.2.13 Performance Evaluation of NCEF

Performance of NCEF should be assessed through a clearly laid-out framework using verifiable indicators. The study focusing on identification and application of such indicators for performance evaluation of the fund presents an analytical structure that can be used to evaluate performance of the fund's portfolio. A list of performance norms that will be necessary for performance evaluation of NCEF is recommended.

Annexure 1
Experts and Stakeholders Consulted During the Course of the Study

S. no.	Workshop 1		Workshop 2	
	Name	Affiliation	Name	Affiliation
1.	M. Govinda Rao	NIPFP, New Delhi	Kirit Parikh, (Panelist and Chair)	IRADe, New Delhi
2.	Rakesh Bhalla (Panelist)	IREDA, New Delhi	Meena Agarwal (Panelist)	Ministry of Finance, New Delhi
3.	Pradeep Dadhich (Panelist)	Deloitte Touche Tohmatsu India, New Delhi	Anil Kumar Jain (Panelist)	Planning Commission, New Delhi
4.	Usha Rao (Panelist)	KFW, New Delhi	D.N. Prasad (Panelist)	Ministry of Coal, New Delhi
5.	Shailly Kedia	TERI, New Delhi	Tapas Sen	NIPFP, New Delhi
6.	Sriya Mohanti	Shakti Foundation, New Delhi	Krishan Dhawan	Shakti Foundation, New Delhi
7.	Sohail Akhtar	Ministry of New Renewable Energy, New Delhi	Shashank Jain	Shakti Foundation, New Delhi
8.	K. Yepthu	IREDA, New Delhi	Chinmaya Kumar Acharya	Shakti Foundation, New Delhi
9.	Aparna Vashisth	TERI, New Delhi	SriyaMohanti	Shakti Foundation, New Delhi
10.	Manish Anand	TERI, New Delhi	AnkitaBhatnagar	Intern, Shakti Foundation, New Delhi
11.	Anandayit Goswami	TERI Africa	ManjushaShukla	IREDA, New Delhi
12.	Anurag Mishra	USAID, New Delhi	Rakesh Bhalla	IREDA, New Delhi
13.	Ashirbad Raha	Climate Group, India	S. Padmanaban	USAID, New Delhi
14.	Pramode Kant	Institute of Green Economy, New Delhi	Anurag Mishra	USAID, New Delhi

R. Pandey et al., *The National Clean Energy Fund of India*, SpringerBriefs in Energy, DOI: 10.1007/978-81-322-1964-4, © The Author(s) 2014

S. no.	Workshop 1		Workshop 2	
	Name	Affiliation	Name	Affiliation
15.	Lydia Powell	Observer Research Foundation, New Delhi	Monali Zeya Hazra	USAID, New Delhi
16.	Karthik Ganeshan	Council on Energy Environment and Water, New Delhi	Mudit Narain	World Bank, New Delhi
17.	Vyoma Jha	Council on Energy Environment and Water, New Delhi	Veena Joshi	Swiss Agency for Development and Cooperation, New Delhi
18.	Rishabh Jain	Council on Energy Environment and Water, New Delhi	PreetiSoni	UNDP, New Delhi
19.	Abhinav Goyal	Centre for Science and Environment, New Delhi	Pramode Kant	Institute of Green Economy, New Delhi
20.	Shradha Kapur	C-Kinetics, New Delhi	Sanjay Dube	Nexant, New Delhi
21.			Anil Kumar	IIEC, New Delhi
22.			Dilip Limaye	SRC Global, Greater Philadelphia Area, USA
23.			Mahesh Patankar	MP Ensystems Advisory Pvt. Ltd. Mumbai
24.			Ahmad Khalid	Adelphi, New Delhi
25.			Manpreet Singh	KPMG, Gurgaon
26.			Stuti Sharma	KPMG, Gurgaon
27.			Sameer Maithel	Greentech Knowledge Solutions Pvt. Ltd., New Delhi
28.			Abhinav Goyal	CSE, New Delhi
29.			Probir Ghosh	Sustainable Resource & Technologies Pvt. Ltd. Colorado, USA
30.			Subir Das	Sustainable Resource & Technologies Pvt. Ltd. Kolkata
31.			Sumana Bhattacharya	Interco-operation India, New Delhi
32.			Bhasker Padigala	WWF, New Delhi

Annexure 2
Norms for Computing Likely Energy Saving from RE Usage

Norms for computing likely annual savings of conventional fuel/electricity through renewable energy deployment

Renewable energy source/system	Likely annual savings of conventional fuel/electricity
Wind power	2.00 million unit/MW
Small hydro power	3.00 million unit/MW
Solar photovoltaic (PV) power	1.66 million unit/MW
Solar PV lantern	50 L K-oil/lantern
Solar PV home lighting system	100 L K-oil/system
Solar thermal energy	1.00 MU/MW
• Power generation	36 TOE/1,000 m^2 collector area
•Thermal energy systems	0.50–0.70 MU/1,000 m^2 collector area
Bio energy	4.00 million unit/MW
1. Bagasse cogeneration	6.00 million unit/MW
2. Biomass power	1,000 TOE/MW_{eq}
3. Biomass energy (thermal)	4.00 million unit/MW
4. Urban and industrial waste to energy	1,000 TOE/MW_{eq}
a. Power generation	450 kg LPG/1,000 m^3 biogas
b. Thermal energy/cogeneration	0.36 million units/1,000 m^3 biogas
5. Family type biogas plants	
6. Medium size biogas plants	

MW Megawatt (installed capacity of power plant)
MW$_{eq}$ Megawatt equivalent-do-
MU Million units (electricity generated saved)
TOE Tonnes of oil equivalent (oil saved)
LPG Liquefied petroleum gas (LPG saved)
Remark 1 Unit of electricity = 0.7 kg of oil

Annexure 3
Templates for Monitoring and Evaluation

Following is a real world example of projects under implementation and recently completed where extensive monitoring and evaluation handles have been provided at the design stage and also being utilised. For purposes of confidentiality the names and agencies have been kept out.

Renewable energy based power generation project: REPP project planning matrix

R. Pandey et al., *The National Clean Energy Fund of India*, SpringerBriefs in Energy,   107
DOI: 10.1007/978-81-322-1964-4, © The Author(s) 2014

Strategy	Indicator	Baseline	Targets	Source of verification	Risks and assumptions
Goal Reduction of greenhouse gas emissions from project in country x's power sector	Cumulative greenhouse gas emission reduction from power generation in project in country x by the end of project (EOP), ktons CO_2	316.4	935.8[a]	DOE records, IPP power generation data, national statistics, national communications, FREPP progress reports and M&E reports	Transparency of decision making, cooperation of stakeholders in the provision of data, stable political environment
Project objective[b] Removal of major barriers to the widespread and cost-effective use of grid-based renewable energy supply via commercially viable renewable energy technologies	Cumulative installed new private sector-owned RE-based power generation capacity by EOP, MW	0[c]	4.7[d]	Survey of IPP investment activities, interviews with prospective investors, FTIB approvals, financing documents from banks/financial institutions, REPP progress reports and M&E reports, FEA annual reports, DOE yearly energy statistics	Investors perceive current governance systems as conducive for doing business in the country. Country's sovereign risk can be managed through government guarantees
	Share of RE in project in country x's power generation mix by EOP, %	52	89.0		
	Cumulative electricity production from RE-based power generation plants by EOP, GWh	494.0	1,505.1[e]		
Outcome 1 Facilitation of investments on energy projects, particularly on RE and biomass based power generation	Cumulative investment on RE-based power generation by EOP, US$ million	0	100	Survey of IPP investment activities, interviews with prospective investors, FTIB approvals, Financing documents from banks/FIs, FREPP progress reports and M&E reports	Investment climate does not deteriorate further, political stability maintained, government able to take on additional contingent liabilities

Strategy	Indicator	Baseline	Targets	Source of verification	Risks and assumptions
Output 1.1: government energy Act	No. of proposed articles on the energy bill that are endorsing RE-based power generation in project in country x	0	Dec 2012	Cabinet records (Gazette)	Relevant key stakeholders such as Prime Minsters Office, Ministry of Justice are supportive and co-operative. Decree based legislation acceptable to foreign investors
	A cabinet-approved comprehensive Energy Act promulgated	0	Dec 2011		
	Institutional reform of DOE to effectively administer the Project in country x Energy legislation	0	Jun 2012	Organisational chart of DOE	Funding for the organisational reforms are approved
Output 1.2: implementing rules and regulations (IRRs)	No. of specific IRRs enforced by EOP	0	Dec 2013	Government gazette, publication of regulation	Political will exists to approve and enforce IRRs. Investors and lenders interests is protected
	No. of revised IRRs proposed to enhance energy act implementation by EOP	0	Dec 2013		
Output 1.3: Government agencies with enhanced regulatory and institutional capacity on energy development, in general, and RE development in particular	No. of RE regulations and legal frameworks administered by DOE senior staff for IPP projects and rural electrification by EOP	0	1	Staff performance annual report, DOE files, license issued	Retention of skilled personnel, job evaluation recommendation
	Percentage of approved RE-based power generation projects that are fully-compliant with DOE-administered RE regulatory and legal frameworks by EOP	0	100		

Strategy	Indicator	Baseline	Targets	Source of verification	Risks and assumptions
Outcome 2					
Technical feasibility of harnessing RE resources are ascertained and made widely known	No. of identified technically viable RE projects EOP	0	6	DOE files, DOE annual reports	FEA is not privatized (sold) completely and co-operates in an effort to mobilize private sector capital.
	No. of investors that made use of available technical information on feasible RE-based energy system projects by EOP	0	20		
Output 2.1: operational centralized energy database system	No. of clients that request services from the central clearinghouse for their RE-based energy systems project EOP	0	300	DOE files, DOE annual reports, information request and feedback forms	Government uses its regulatory powers to override commercial confidentiality concerns and orders data and information to be made public
	No. of clients that make use of the central energy database system each year	0	150		
	Percentage of clearinghouse and central energy database system clients each year that are satisfied with the services received	0	80		
	No. of implemented RE-based power generation projects that were facilitated by the central clearing house system by EOP	0	20		
Output 2.2: completed and published RE resource assessments	No. of comprehensive RE resource assessments completed by EOP	0	12	DOE files, DOE annual reports, resource assessment reports, Met and PWD annual reports, forestry annual reports, FEA annual reports, IPP proposals	Resource data and assessments are accurate and reliable
	Average percentage increase in currently known RE potentials that was established after the RE resource assessments	0	Dec 2013		
	No. of investors that made use of the RE resource assessment data/information in the design of their RE-based power generation projects by EOP	0	6		

Strategy	Indicator	Baseline	Targets	Source of verification	Risks and assumptions
Output 2.3: assessed feasibility of RE investments	No. of completed and published new feasibility studies of IPP investments by EOP	0	6	DOE website, IPP proposals, DOE Corporate plan, FEA Annual reports, FTIB records	Co-financing components secured
	No. of planned new feasibility analyses to be carried out (after FREPP) by EOP	0	4		
	Percentage of interested investors in project in country x that expressed confidence in the technical and financial viabilities of RE-based power generation projects by EOP	0	30		
Outcome 3 Markets for specific renewable energy technologies are supported	No. of additional rural households that have access to green electricity by EOP	0	10,000	Signed PPA, contracts, financing agreements, shareholder agreements; HIES report, DOE annual reports and records	Political will exists to allow private sector investment in IPP. FEA remains a state owned utility, Private sector investors perceive Project in country x as viable destination. Government ruling in other sectors does not undermine investors' confidence
	No. of financial closures achieved for new RE-based power generation projects by EOP	0	20		
	No. of RET system equipment/component suppliers & distributors in Project in country x by EOP	5	7		
	Overall volume of business in the RE market in Project in country x by EOP, US$ million	0	100		
Output 3.1: designed and implemented RE-based power generation demonstration	Overall installed capacity of RE-based power generation demo projects by EOP, MW	0	4.7	DOE records, DOE Annual Report, FEA Annual Reports, FEA Corporate plan, DOE Corporate plan	Political will including Government funding available to implement demo projects, resource ownership not an issue, demo projects are sucessful including no major natural disaster
	No. of demo projects that are both operationally and financially viable by EOP	0	10		
	No. of planned RE-based power generation projects that are replicating any of the demo projects by EOP	1	16		
	Total installed capacity of replication RE-based power generation projects by EOP	0	At least 3		

Strategy	Indicator	Baseline	Targets	Source of verification	Risks and assumptions
Output 3.2: prepared standard power purchase agreement (PPA) for IPPs	Endorsed Standard Power Purchase Agreement (SPPA) templates that are used for IPP projects in project in country x	0	1[f]	Tender documents for competitive IPP procurement, DOE website	Interests of all parties are adequately protected
	No. of IPP RE-based power projects that made use of any of the approved SPPA templates by EOP	0	6		
Output 3.3: completed investment promotion package	No. of prospective investors making enquiries with government agencies	0	15	DOE Annual Report, Record of Investor proposals, Investment Forum Report, List of Participants	Role of DOE as an investment facilitator accepted by Cabinet
	Cumulative number of investors that expressed and planned to invest and implement RE-based power generation projects by EOP	0	10		
Output 3.4: Completed assessment and developed RE incentives schemes	A comprehensive report on options and issues related to the establishment of a subsidy fund for private sector renewable energy investment published	0[g]	Jun 2012	Completed feasibility report	NB: Ref RE fund design through RESCO
Outcome 4 Renewable energy developments integrated into national energy plan towards 100 % electrification of project in country x	Cabinet approved-electrification master plan	0	Dec 2013	Master plan document, DOE webpage, DOE files, FEA annual report, DOE energy statistics year book	The goal of 100 % electrification is maintained; renewable based power generation continue to be cost competitive; renewable energy potential is sufficient to meet current and future demand
	Average annual budget for the electrification master plan by EOP, US$ million	0	10		
	Percentage utilization of country's RE resources (for power purposes) by EOP	52	90		

Strategy	Indicator	Baseline	Targets	Source of verification	Risks and assumptions
Output 4.1: completed training programme on integrated energy planning (IEP) and administrative energy policy for government personnel	No. of GOF personnel trained on IEP and energy policy each year starting year 2011	2	6	DOE records, training assessment and training plan document, records of training sessions, FNU and USP annual faculty reports, certificates	Training institutions are willing to co-operate with government, qualified staff can be retained within the energy sector
	Percentage trained GOF personnel that are actively engaged in RE-based power generation policy making, planning and implementation, operations and evaluation by EOP	0	50		
	No. of training institutions that are capable and qualified in IEP and energy policy training/capacity building by EOP	2	2		
Output 4.2: completed and approved National Electrification Master Plan	Cabinet approved-Electrification Master plan	0	Dec 2013	Master plan document, DOE webpage, DOE files, FEA annual report, DOE energy statistics year book	The goal of 100 % electrification is maintained
	Average annual budget for the electrification master plan by EOP, US$ million	0	10		

[a]Minimum end-of-project CO_2 emission reduction from demonstrations only (3.2 MW VRE PP, and 25 % of biofuel mills operational by EOP)

[b]Objective monitored quarterly ERBM and annually in project implementation review

[c]Considering that FSC and Tropik Woods are not entirely IPPs

[d]This is minimum, taking consideration only of the 3.2 MW VRE biomass-based power plant and 5 × 300 kW diesel engines using biodiesel produced by 5 biofuel mills

[e]This is minimum, taking consideration of baseline RE electricity + electricity generation only from VRE biomass-based PP and 5 biodiesel power generation units

[f]There will be only 1 standard template since there is only 1 transmission and distribution utility

[g]Prospective private RE investors do not commit funds, as investments are commercially not viable without support. Country X's renewable energy industry remains small and weak. RE investment remains dependent on donor funding

Field visit-monitoring and evaluation small grantees' projects

S. No.			Information
	Field Visit- Monitoring & Evaluation Small Grantees' Projects		
1.	Title of the project		
2.	Name of the lead grantee & address		
3.	Name & address of the partner organization(s)		
4.	Type of the Project		
	▪ Desk Research ▪ Training & capacity building ▪ Information dissemination / outreach ▪ Field/ Technology Demonstration		
5.	Project location		
6.	Start date of the project		
7.	End date of the project		
8.	A brief description of the project		
9.	Project milestones & time frame		

S. No.	Milestone	Date as per the contract		Actual	Status
1.					
2.					
3.					
4.					

10.	Project deliverables & time frame		

S. No.	Deliverables	Completion date as per contract		Status
1.				
2.				
3.				
4.				

11.	Key observations from the field visit		
12.	Major deviations, if any, from project scope, methodology, time frame		
13.	Challenges faced / problems encountered & strategy		
14.	Information dissemination / outreach activities for the project		
15.	Any assistance required from USAID/WI/WII		

16.	Document Set	Type of report	Number expected	Number received
		Monthly Reports		
		Quarterly Reports		
		Quarterly Financial Report		
		Closure/Final Reports